Southeast Asia Workshop on Aerospace Engineering (SAWAE 2019)

IOP Conference Series: Materials Science and Engineering
Volume 642

Kuala Lumpur, Malaysia
29 – 30 August 2019

ISBN: 978-1-5108-9842-4
ISSN: 1757-8981

Printed from e-media with permission by:

Curran Associates, Inc.
57 Morehouse Lane
Red Hook, NY 12571

Some format issues inherent in the e-media version may also appear in this print version.

Institute of Physics
Dirac House, Temple Back
Bristol BS1 6BE UK

Phone: 44 1 17 929 7481
Fax: 44 1 17 920 0979

techtracking@iop.org

Additional copies of this publication are available from:

Curran Associates, Inc.
57 Morehouse Lane
Red Hook, NY 12571 USA
Phone: 845-758-0400
Fax: 845-758-2633
Email: curran@proceedings.com
Web: www.proceedings.com

Table of contents

Volume 642

Southeast Asia Workshop on Aerospace Engineering (SAWAE 2019)

29–30 August 2019, Kuala Lumpur, Malaysia

Accepted papers received: 10 September 2019
Published online: 24 October 2019

Preface

Papers

The Southeast Asia Workshop on Aerospace Engineering (SAWAE) was founded in 2016 and first hosted by National Cheng Kung University, Taiwan. It is organized in conjunction with the annual meeting of Southeast Asia Network in Aerospace Engineering (SANAE) and is hosted in turns by the constitutional members of SANAE. Since then, subsequent SAWAE events were hosted by Ho Chi Minh City University of Technology, Vietnam (2017) and Kasetsart University, Thailand (2018). For this fourth edition of the workshop, it is jointly organized by Universiti Putra Malaysia and Universiti Teknologi Malaysia, which also marks the first time it is held in Malaysia.

The primary aim of SAWAE is to provide an effective regional platform to discuss and exchange recent developments, new ideas, and experiences in the field of aerospace engineering between academia, government agencies, and industries in the region. It is hoped that this event will help and facilitate close dialogues, networking, and collaborations among experts on issues related to research and technological development in aerospace engineering, as well as in human resource development and education. This is directly aligned with the selected theme for this edition, which is "Aerospace Research and Development: Challenges and Opportunities".

IOP Conf. Series: Materials Science and Engineering **642** (2019) 011002 doi:10.1088/1757-899X/642/1/011002

Peer review statement

All papers published in this volume of *IOP Conference Series: Materials Science and Engineering* have been peer reviewed through processes administered by the proceedings Editors. Reviews were conducted by expert referees to the professional and scientific standards expected of a proceedings journal published by IOP Publishing.

Case Study on Aerodynamics Stability of Bixel Wing-In-Ground Effect Craft

Afifi.A Pua'at[1] , Zulkifli Matdaud[1], Amzari Zhahir[1]*, Mohamed Tarmizi Ahmad[2] and Aziz Hassan[3]

[1]Dept. of Aerospace Eng., Universiti Putra Malaysia, 43400 Serdang, Malaysia.
[2]Assoc. Professor, Dept. of Mechanical Eng., Universiti Pertahanan Nasional Malaysia, Kuala Lumpur, Malaysia.
[3]Retired Professor in Mechanical Engineering (Automotive), Universiti Teknologi Malaysia.

*Corresponding author e-mail: amzari@upm.edu.my

Abstract. Wing-in-ground effect (WIG) crafts fly efficiently by exploiting the ground effect i.e. generating a notable rise in the lift-over-drag ratio when a three-dimensional wing approaches a large flat plane either water or ground surface. A successful WIG craft will create another mode of transportation categorise in between commercial airplanes and marine vessels. Hence, this paper presents a case study of Bixel Wing-In-Ground-Effect aerodynamic (Bixel WIG). The initial task is to simulate Bixel WIG aerodynamic using XFLR5 software. The craft configuration is a double wing flat aerofoil wave rider design with aspect ratio variation from 0.25 to 0.5 as in USA PATENT #5105898 C.BIXEL. Based on the result of the simulation, the design of the six seater double wing Bixel WIG is proposed. The design is benchmarked against Lippisch RHB X-114 and UH-18 SPW.

Keywords: wing-in-ground-effect, WIGE craft, Aerodynamics, Bixel, hoverwing

1. Introduction

A phenomenon known as ground effect occurs, when a three-dimensional wing flies closes to a large flat plane either water or ground surface. This result is in a notable rise in lift-to-drag ratio. In this paper, the authors would like to introduce a patented Double wing flat aerofoil design from Charles G. Bixel known as BIXEL WIG craft.

BIXEL WIG craft is a sea going vessel that exploits the wing-in-ground effect phenomenon by flying at low altitude relative to the water surface. The low aspect ratio and highly elongated flat wing provides a travelling capability within the ground effect region at a safe distance over the water [1]. BIXEL WIG craft has a high Reynold number although it operates at low cruising speed (max 150 km/h). The construction cost of BIXEL WIG craft is expected to be less than that of an airplane of the same load capacity. BIXEL WIG crafts are categorised as a ship and operates under the International Maritime Organization (IMO), hence the design requirements and regulations for BIXEL WIG crafts are not as rigorous as those governing aircraft industry. The thick wing and low aspect ratio of BIXEL WIG craft are simple to assemble because it is operating at a very low altitude as compared to the aircraft or seaplanes.

Although there are various types of WIG crafts that have been designed and developed globally, they have not been successfully commercialised. One major reason is the uncertainty of the performance of the vessel in the open sea and the service operation costs. The operating cost of the vessel increase as it operates on the sea during take-off which requires higher power to overcome the hydrodynamic drags. After take-off and during cruising, less power is required hence an inefficient engine capacity usage. Reducing the power requirement during take-off is the key to successful development of a BIXEL WIG craft. Therefore, the authors have been studying the BIXEL configuration that has shown potential solution to the commercial viability.

Section 2 of this paper reports the aerodynamic characteristic of Bixel WIG craft. Section 3 focuses the XFLR5 simulation method used to obtain the preliminary data is described. Section 4 analyses the aerodynamic data obtain based on the double wing flat aerofoil wave rider design with aspect ratio variation from 0.25 to 0.5 as in USA PATENT #5105898 C.BIXEL. Section 5 proposes six seater hoverwing configurations, and finally concluding remarks are in Section 6.

2. Aerodynamic characteristics of Bixel Wing-In-Ground effect craft

For a start, an explanation of a generic WIG craft need to be described before further elaboration on the specific Bixel WIG craft. Generally, a WIG craft is an aircraft operating in close proximity of the ground or any level surfaces [2]. Under the International Maritime Organization (IMO), it is classified as a marine vessel [3]. The technology was develop by the Russian military for transporting heavy military equipment. The iconic WIG known to the west is 'Caspian Sea Monster' [4]. Later the German develop a smaller WIG for commercial purposes [5]. Currently, beside the above two countries China and Korea are actively pursuing the development of passenger WIG [6-7]. China (DXF100-Tianyi-1), Korea (C&S AMT Aron7), Iran (Bavar 2) and Singapore (AF8 001-Airfish) are working from small to large capacity WIG crafts. The major difficulties encountered is due to the vessel operating in three different environment; on land or water, ground effect zone and outside ground effect (free stream) [8]. Each of this environment require different vehicle characteristic for stability and efficiency. Improving the aerodynamic performances of WIG is a challenge. This is due to the limited published information on the performance and the commercial viability. Most of the limited information are based on simulations or performance of remote control (RC) model.

The Bixel WIG configuration is different from the current design by the Russian and the German. This type of WIG craft has a double wing configurations augmented by the hover craft capabilities for acceleration to take-off speed. It is claimed that the Bixel WIG craft has is more efficient than modern aircraft by 250%, faster than cargo ships by 15 folds and able to fly over various terrains (land and sea) [1]. Theoretically, the integration of wing and fuselage will increase lift-to-drag ratio since the lift area increases due to fuselage turning into wing. Furthermore the rectangular shape wing also contributes to the lift.

In this paper, a first order analysis of aerodynamic simulation on Bixel WIG craft is done by using XFLR5 software [9]. Aerodynamic characteristic such as lift, drag, pitching moment and aerodynamic centre are obtained to guide the design process.

3. Bixel WIG craft: Aerodynamic Investigation

In order to justify that WIG craft is totally different from standard aircraft, it is important to first outline a few aerodynamic concepts. Theoretically, the raise of lift to drag ratio represents the raise of body efficiency [10]. As a wing moves through air a resultant force is generated. This resultant force can be express into lift and induced drag component. Another type of drag is referred to as parasitic drag. This type of drag is the result of friction as the aircraft moves through the air. The combination of parasitic drag and induced drag is called total drag, as shown in Fig. 1.

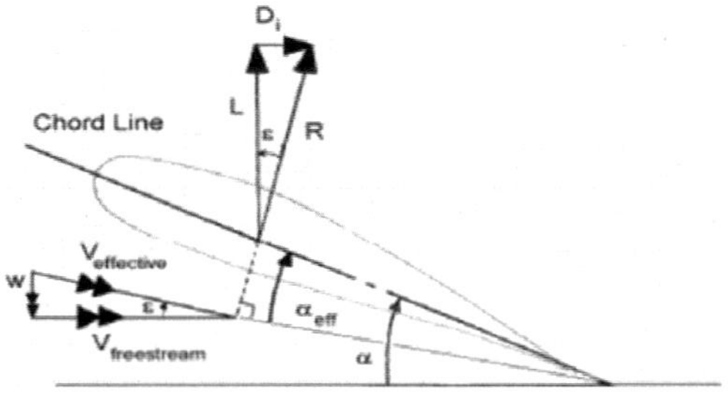

Fig. 1: Lift and drag of an aerofoil [11]

Another important point to highlight is the ground effect. This effect occurs by the presence of a physical boundary at short distances under the wing. The effect is inversely proportional to the distance between the boundary and the wing. Since the lift to drag ratio is the wing efficiency, the ground effect contributed significant increase to the wing efficiency.

For flow over a smooth flat plate, the transition from laminar to turbulent starts roughly at about Re 1×10^5. The flow becomes fully turbulent when Reynolds number reaches around 3×10^6 [12]. Generally, an accepted value for low Reynolds number is below 3×10^6. The equation (1) for Reynold number is as stated below:

$$\mathrm{Re} = \frac{\rho V x}{\mu} = \frac{V x}{\nu} \qquad (1)$$

where V is the upstream velocity, ρ is the density of the fluid, μ is the dynamic viscosity of the fluid, ν is the kinematic viscosity of the fluid and x is the characteristic length of the geometry, which, for a flat plate, is the length of the plate in the flow direction.

The Bixel configuration model as stated in [1] is a double wing flat aerofoil wave rider design with aspect ratio variation from 0.25 to 0.5 as in USA PATENT #5105898 C.BIXEL was simulated and its' aerodynamic characteristic such as coefficient of lift (C_l), coefficient of drag (C_d), angle of attack (α), and coefficient of moment (C_m) are being drawn as first order analysis. Noted that the ground effect is also applied in the analysis with the height from ground to underbody surface is set to be maximum of 500 mm. (Fig. 2 illustrates an artist view of the conceptual Bixel WIG craft). The Bixel WIG craft model dimension is about 332 mm x 500 mm with wing and body act as one big wing. The investigation is done to define the aerodynamic capability of this type of configuration.

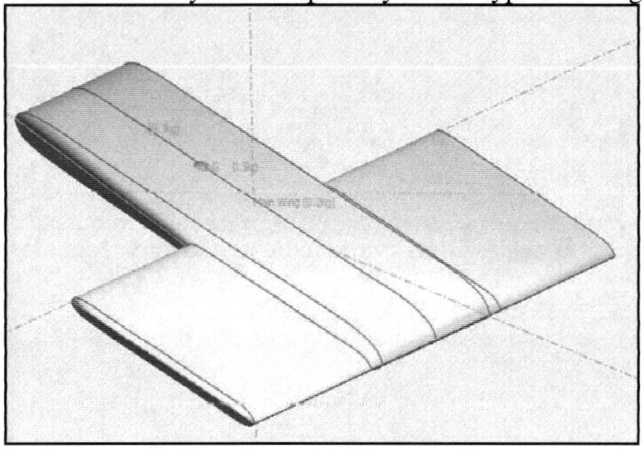

Fig. 2: Bixel WIG craft model illustrated in XFLR5 (ISO)

IOP Conf. Series: Materials Science and Engineering **642** (2019) 012001 doi:10.1088/1757-899X/642/1/012001

4. Analysis

4.1. Analysis software and result

The chosen software analysis to produce preliminary aerodynamic result on this Bixel WIG craft model is XFLR5. is a free aerofoil, wing and aircraft analysis instrument that operates at low Reynolds numbers. Fig. 3 to Fig. 4 shows the illustrated Bixel WIG craft configuration based on the previous scale model in (3). The several test was simulated for dimensional variation as the percentage of vertical height (H) divided by wing span (WS) which is 20%WS (0.1m), 40%WS (0.2m), 60%WS (0.3m), 80%WS (0.4), and 100%WS (0.5m). The velocities variation applied is 10 m/s and 25 m/s. The analysis uses the Vortex lattice method (VLM) and the maximum height above the surface applied is 500mm. Fig. 5 to Fig. 7 shows the various aerodynamic properties as the final output.

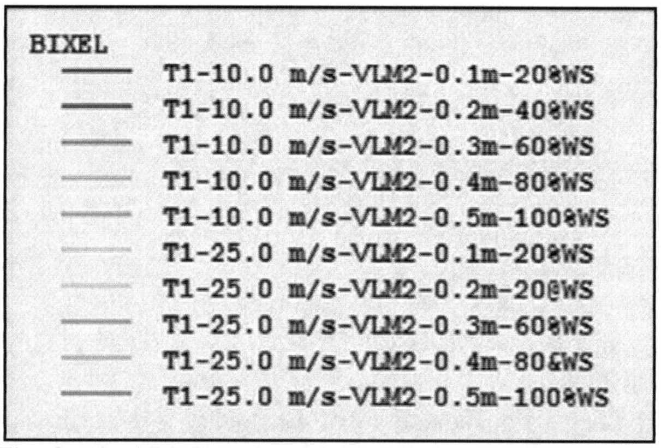

Fig. 3: Bixel WIG craft model height variations in XFLR5

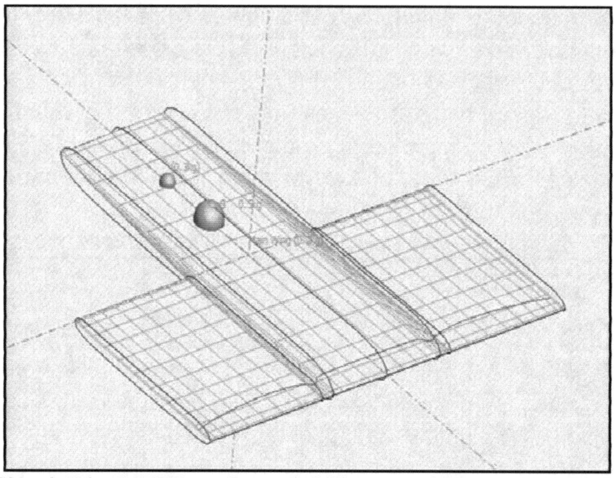

Fig. 4: Bixel WIG craft model illustrated in XFLR5 (ISO)

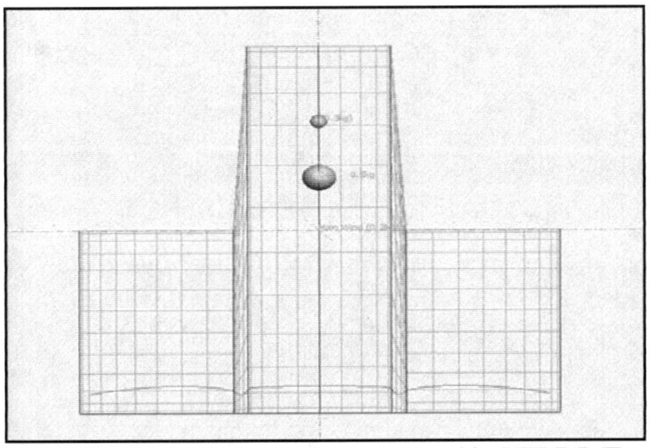

Fig. 5: Bixel WIG craft model illustrated in XFLR5 (Top)

4.1.1.Lift and drag

Fig. 7 shows that the highest C_l achieved is 0.456, $\alpha = 9.822$ with 20%WS at both 10 m/s and 25 m/s velocities. This indicates that as the angle of attack (α) increases, the coefficient of lift (C_l) also increased. The amount of lift is increase proportionally to model α. Fig. 8 shows that at 10 m/s, C_d = 0.018, $\alpha = 0.011$, and at 25 m/s. C_d = 0.014, α =0.022. The amount of drag is existed even when the model α is at 0. Fig. 8 also showed that less drag is achieved as the model gain higher velocity. Fig. 9 shows the $C_l - C_d$ variation for all vertical height (H) at both speeds. The amount of C_l increases as the C_d increase.

4.1.2.Aerodynamic efficiency

The amount of efficiency C_l/C_d is increasing proportionally with the rises of the model α in all velocities as illustrated in Fig. 9.

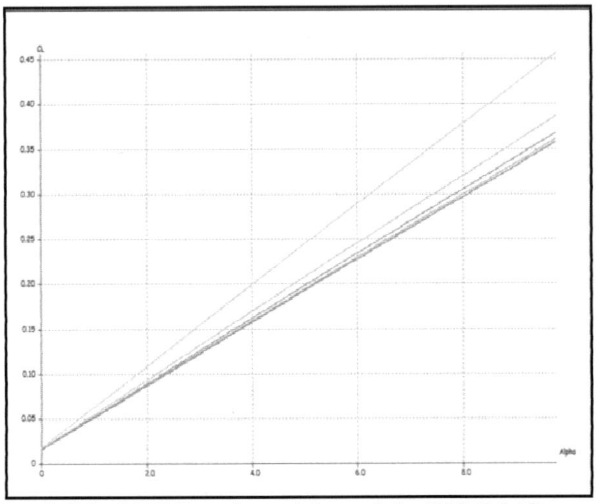

Fig. 6: Bixel WIG craft model result in XFLR5 (C_L vs α)

IOP Conf. Series: Materials Science and Engineering **642** (2019) 012001 doi:10.1088/1757-899X/642/1/012001

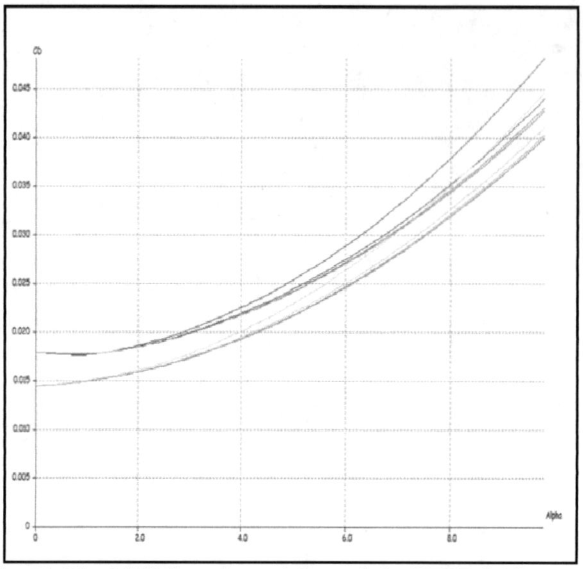

Fig. 7: Bixel WIG craft model result in XFLR5 (C_D vs α)

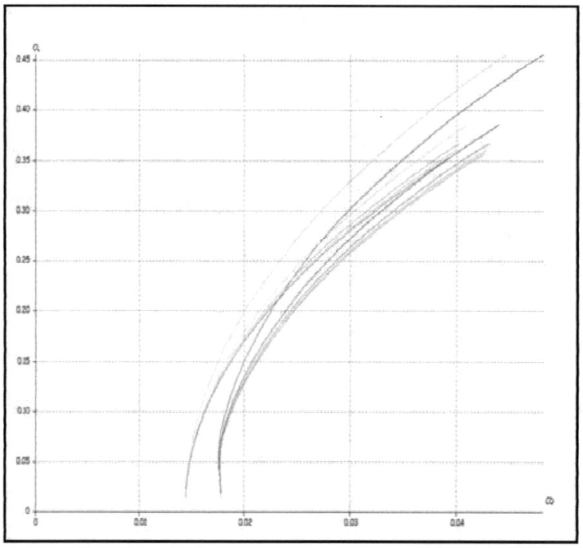

Fig. 8: Bixel WIG craft model result in XFLR5 (C_L vs C_D)

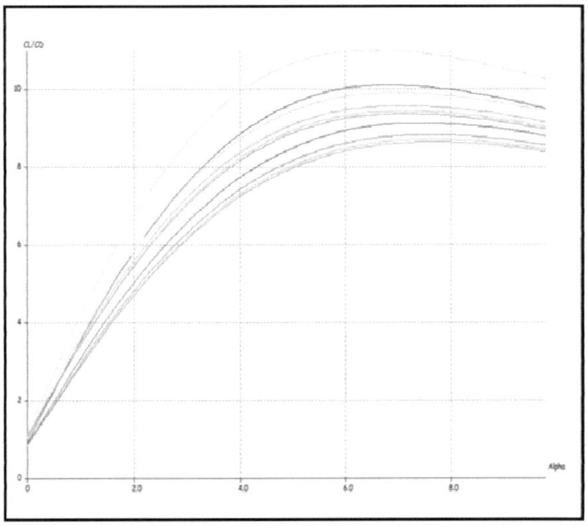

Fig. 9: Bixel WIG craft model result in XFLR5 (C_L/C_D vs α)

4.1.3. Pitching moment

Pitching moments C_m against the angle of attack α for the Bixel WIG craft model is illustrated in Fig. 10. From the C_m - α curve, the model has longitudinal static stability for all vertical height (H) and both speed. Normally, most of the WIG craft's operational cruise condition is within the main plane attack angle of 2° to 4° [13] and in this graph, the negative slope for positive α indicates stability in pitching moment. The graph indicates positive pitching moments in this range. By making an assumption of setting $\alpha = 4°$ for cruising condition, increasing the aerodynamic efficiency will potentially rise 40% of its' overall lift for the current type of WIG. A continuous trimming on the elevators' pitching moments need to be done in order for Bixel WIG crafts to operate in this condition. Continuous flight control can be used to reduce the workload of the flying crew as well as to automate and manage many related tasks [14]. Stability and control understanding becomes important because it is hard to master load and vertical controls [15].

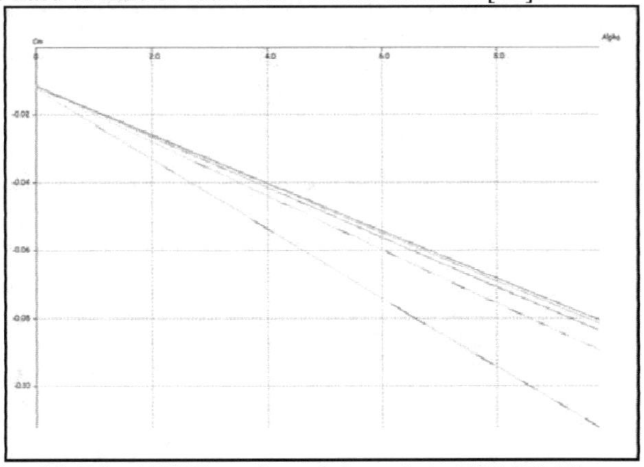

Fig. 10: Bixel WIG craft model result in XFLR5 (C_M vs α)

4.1.4. Pressure centre line variant

Besides the aerodynamic features of a full model, the experiments also went on the determination of the pressure center line pattern acting below the model wing and body surfaces. [16]. By using XFLR5, the pressure centre line pattern is as shown in Fig. 11 to Fig. 14.

IOP Conf. Series: Materials Science and Engineering **642** (2019) 012001 doi:10.1088/1757-899X/642/1/012001

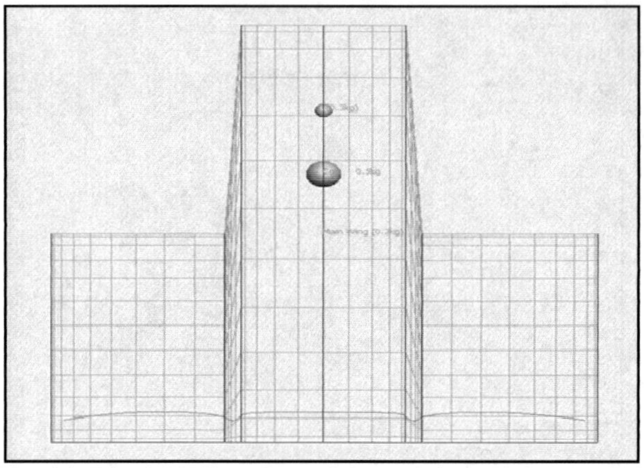

Fig. 11: Bixel WIG craft model result in XFLR5 ($\alpha = 0°$)

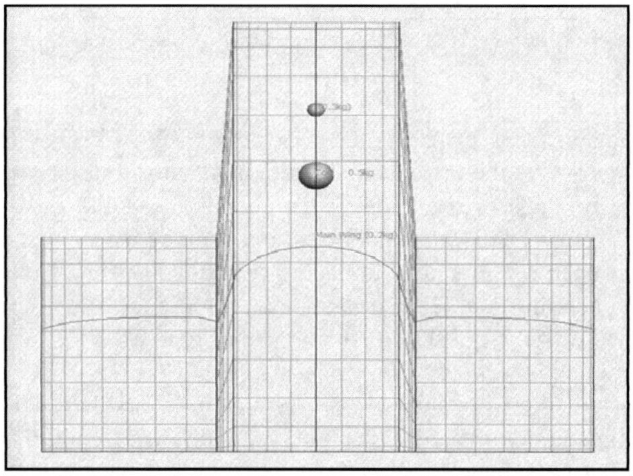

Fig. 12: Bixel WIG craft model result in XFLR5 ($\alpha = 1°$)

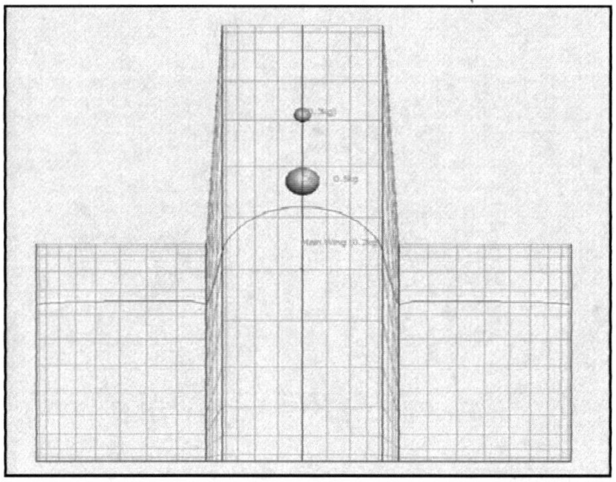

Fig. 13: Bixel WIG craft model result in XFLR5 ($\alpha = 3°$)

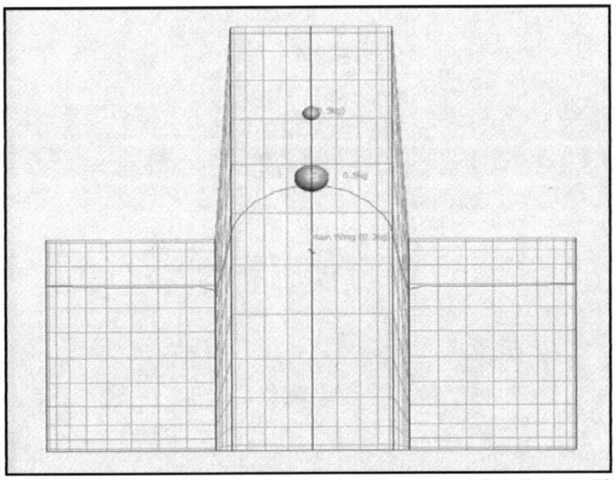

Fig. 14: Bixel WIG craft model result in XFLR5 (α = 7°)

Fig. 10 to Fig. 13 indicates that higher lift force acting on the model middle surfaces. This is due to the increases of lift area compare to both left and right wing [17]. As angle of attack increases more than α = 3 degree, the lift line acts on the wing and body surfaces is almost in the same location. The graph also showed a significant rises of pressure centre line location on the body surfaces from α = 0° to α = 1° [18].

From result showed in section 3.1.4, Bixel type of wing shows a significant location different in lift pattern from $\alpha = 0$ ° to $\alpha = 1$ ° degree. Whereas the lift pattern showed a steady location different starting from $\alpha = 1$°. Fig. 15, Fig. 16 and Fig. 17 show the different of α change as stated before.

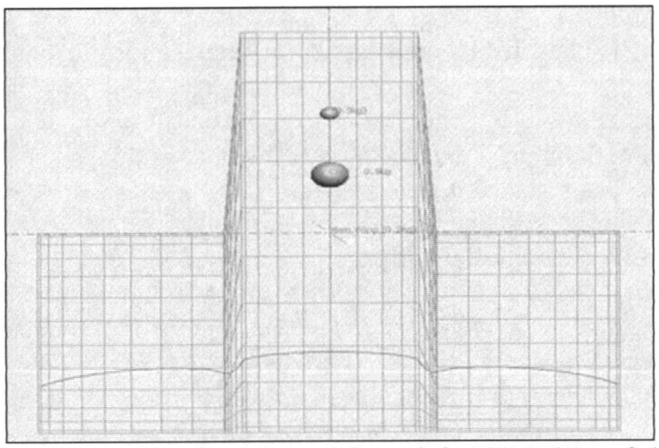

Fig. 15: Bixel WIG craft model result in XFLR5 (α = 0.1°)

IOP Conf. Series: Materials Science and Engineering **642** (2019) 012001 doi:10.1088/1757-899X/642/1/012001

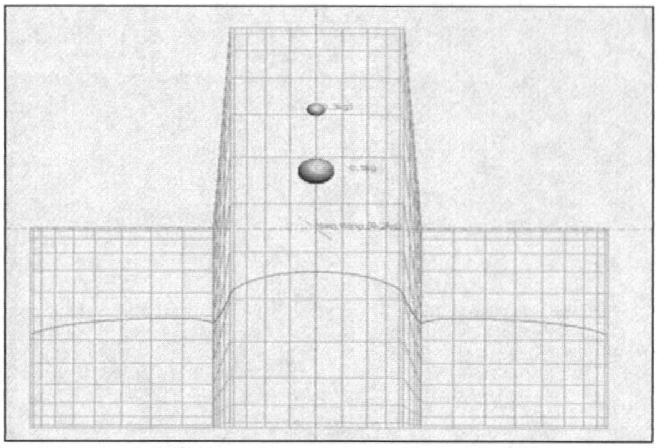

Fig. 16: Bixel WIG craft model result in XFLR5 ($\alpha = 0.5°$)

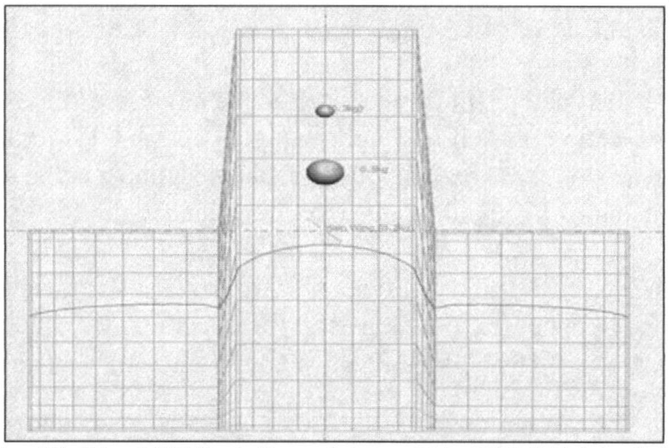

Fig. 17: Bixel WIG craft model result in XFLR5 ($\alpha = 0.9°$)

Based on tests and calculation, it is necessary to validate all aerodynamic analysis assumptions [19]. For that reason the authors proposed to initiate flight testing with a remote control of a small scale Bixel WIG craft.

5. Proposed six seater Hovering Configuration

The X-112, one of the first WIG craft was developed by Alexander Lippisch, in 1963. The X-112 planform was a reverse delta wing with low aspect ratio, anhedral and forward sweep [20]. This configuration is now referred to as the Lippisch planform. The reversed delta planform of Lippisch is said [21] to have a lower movement of the centre of pressure while obtaining a high lift to drag ratio [22]. The planform results in a lesser change in pitch stability and thus has a reduced tail area when comparing with ram wing craft [23]. Due to the configuration of the planform, it is less significant for the change in pitching moment with height above ground plan. This decreases the control power needed for height transition and thus decreases the area of the tail plane.

Fig. 18: Lippisch X-114 [24]

The subsequent development of X-112 is the X-114 Hoverwing as shown in Fig. 18. It is designed as a high-performance hovercraft capable of flying at speeds exceeding 75 mph over rough water. Cruise altitude is 2 to 6 feet from ground level and the WIG craft can climb up to 20 feet to clear large obstacles. Even though WIG craft operates in the ground effect region, is categorised as a boat and thus does not require a pilot's license. This in turn brings a wide range of new opportunities to the commercial and tourism industry. Recent development by Universal Hovercraft is the UH18SPW as showed in Fig. 19 and further develops into UH19XRW.

Fig. 19: UH18SPW Hoverwing [24]

Fig. 20: Small scale Bixel WIG craft model (ISO)

The research is based on the UH18SPW platform which is modified to incorporate the Bixel platform configuration to accommodate six passengers as in X-114 [25]. The proposed Bixel WIG craft mock up model platform is as shown in Fig. 20.

Fig. 21: Small scale Bixel WIG craft model (Side ISO)

Fig. 22: Small scale Bixel WIG craft model (Front ISO)

As a final result the authors concluded that the proposed Bixel type of WIG craft is able to produce good lift and it has longitudinal static stability.

6. Conclusion

Based on the analysis, the proposed Bixel type wing configurations are able to achieved higher lift for a given wingspan [26]. For this type of configurations, it is shown that the longitudinal static stability can be achieve due to its' negative slope of the pitching moment plot.

The performance analysis described in the previous section has not been validated. In addition, further analysis will be carry out on the modified basic rectangular wing configuration used in UH18SPW by incorporating endplate and later the addition winglet [27].

Acknowledgement
The author would like to thank the Department of Aerospace Engineering UPM for providing financial support for the preparation of the article in the writing of the report and in the decision to submit the paper for publication.

References

[1] Charles.G, Bixel Jr. Hovercraft ground-effect vehicle. USA Patent #5105898 C.BIXEL, Apr. 21, 1992: 1-11.
[2] Reeves JML. The case for surface effect research, platform applications and development opportunities. NATO-A-GARD fluid mechanic panel (FMP)symposium in long range and long range endurance operation of aircraft 1993; session 1A, paper no 4: 24-27.
[3] Bogdanov AI. The Problems of Ekranoplan Certification Conception and Development of IMO Safety Development. In: Proceeding of international workshop ekranoplans- flying ships of the 21st century. University of New South Wales 1995: 128-147.
[4] Belavin NI. Letauschie Suda-ekranoplany. Katera I Yakhty 1968; 15.
[5] Fisher H, Matjasic K. From Airfish to Hoverwing. In:Proceeding of the international workshop WISE up to ekranoplan GEMs 1998: 69-89.
[6] Ming LS, Xai LK, Hua YC, Bo N, Nai YX. Development of wing in ground effect at CSSRC. In: Proceeding of the international workshop Ekranoplans and very fast craft 1989: 244.
[7] Kang CG. Introduction to large WIG craft project. In: Proceeding of the international symposium on WIG craft 2005: 77-84.
[8] Hooker S. A review of current technical knowledge necessary to build large scale wing in surface effect craft. In: Intersociety advanced marine vehicles conference 1989: 367-429.
[9] J. Jensin Joshua,P. Azhaguraja, P. Banumathi, P. Abirami. Lateral and Longitudinal Stability Analysis of UAV Using XFLR5. Department of Aeronautical Engineering, Adhiyamaan College of Enginering,Hosur, WA 2016: 2.
[10] Cheong Ming Yin, Surjatin Wiriadidjaja, Dayang Laila Abd Majid, Fairuz Izzudin bin Romli, Azmin Shakrine, Mohd Rafie and Amzari Zhahir. Review on The Cost And Performance of A Wige Craft: A Commercialization Prospective. International Journal of Engineering and Technology 2017; 7: 2.
[11] Eng. Islam Ahmed Amin. Efficiency of Wing-In-Ground Marine Crafts. Faculty of Engineering, Port Said University 2006: 144.
[12] Collinson, R.P.G. Introduction to Avionics. Springer 1996.
[13] Wiriadidjaja.S, Mohamad.Z.H, and Rafie.A.S.M. Wing-In-Ground-Effect Craft as a Potential Domestic Transport Vehicle. Universiti Putra Malaysia 2017: 3.
[14] Kirillovykh VN. Russian ekranoplan. In: Proceeding of the international workshop on twenty first century flying ships 1995: 71-177.
[15] Yunus.A.C,John.M.C. Fluid Mechanics: Fundamental and Application. McGrew-Hill Series 2006: 11-15.
[16] Christos D. Flight Dynamics of a Wing in Ground Effect Air Vehicle utilizing a RAM Air Wing. Cranfield University 2006: 133.
[17] Wiriadidjaja.S, Zhahir.A, Mohamad Z.H, Razali.S, Puaat.A.A, and Ahmad.M.T. Wing In Ground Effect Craft: a Case Study in Aerodynamics. International Journal of Engineering and Technology 2017: 5-9: 4.
[18] Michael Halloran, Sean O'Meara. Wing in Ground Effect Craft Review. Aeronautical and Maritime Research Laboratory 1999: 21
[19] Juhee Lee,Chang-suk Han, and Chang-Hwan Bae. Influence of Wing Configurations on Aerodynamic Characteristics of Wings in Ground Effect. Hoseo University, Asan Chungnam 2010: pp 10.
[20] Kyoungwoo Park, Chol Ho Hong, Kwang Soo Kim and Juhee Lee. Effect of Endplate Shape on Performance and Stability of Wings-in Ground (WIG) Craft. World Academy of Science, Engineering and Technology 2008: pp 4.
[21] Dorian N. and Draghita I. The Aerodynamic Force Calculus for a Plate Immersed in a Uniform Air Stream Using SolidWorks Flow Simulation Module. Department of Materials Engineering, Mechanical and Industrial 2011: 5.
[22] Bill Husa, WIG Configuration Development From Component Matrix. Orion Technologies,

Snohomish 2000: 4.

[23] Zhigang Yang, Wei Yang and Yulong Li. Analysis of Two Configurations for a Commercial WIG Craft Based on CFD. Shanghai Automotive Wind Tunnel Center, Tongji University, Shanghai 2009: 5.

[24] Bleischwitz R. de Kat and B. Ganapathisubramani. Aeromechanics of membrane and rigid wings in and out of ground-effect at moderate Reynolds Numbers. Aerodynamics & Flight Mechanics Group 2016: 15.

[25] Nebylov A.V. Controlled Flight Close To Rough Sea: Strategies and Means. State University of Aerospace Instrumentation 2002: 3.

[26] Akagi S. Synthetic Aspects of Transport Economy and Transport Vehicle Performance with Reference To High Speed Marine Vehicles. FAST 91, Trondheim, Norway 1991: 277.

[27] Julien Alexandre dit Sandretto, Douglas Picciani de Suza, and Alexandre Chapoutot. Appropriate Design Guided by Simulation: An Hovercraft Application. ENSTA ParisTech 2016: 7.

Experimental Study on the Effect of Two Equal-Sized Holes Lying at Different Angles in a CFRP Plate under Bending

Phacharaporn Bunyawanichakul[1]*, Paiboon Limpitipanich[2], and Pongwit Siribodhi[1]

[1]Design Clinic Research Unit, Department of Aerospace Engineering, Faculty of Engineering, Kasetsart University, 50 Ngam Wong Wan Rd., Ladyaow, Chatuchak, Bangkok, 10900, Thailand.
[2]Department of Mechanical Engineering, Faculty of Engineering, Burapha University, 169 Longhard-bangsaen Rd., Saensook, Muang, Chonburi, 20131, Thailand.

*Corresponding author e-mail:phacharaporn.b@ku.ac.th

Abstract. A bending test method was designed to investigate the failure characteristics of drilling carbon-fiber reinforced composite (CFRP) plates. The CFRP plates were fabricated and drilled to join other parts for use in aircraft structures (riveting or bolting). Each sample had two equal-sized holes at 0, θ, or -θ degrees at a fixed distance of 36 mm in the longitudinal direction. The sample dimensions were identified following a study on tensile testing single-holed samples based on the ASTM D5961 standard. The tensile test focused on three cases of laminate: $[0]_{14}$, $[22.5]_{14}$, and $[45]_{14}$. The maximum mean forces for $[0]_{14}$, $[22.5]_{14}$, and $[45]_{14}$ laminate configurations were 6.32, 6.59, and 7.12 kN respectively. These results were implemented into a bending test which included two configurations of laminate: $[0]_{14}$ and $[45]_{14}$ and two equal-sized holes with a variable distance of 0 and 9 mm in the transverse direction of the plate. The bending test configuration is presented in this paper. The results revealed different load-bearing capabilities in each of the three types of specimen configuration. Samples with two holes lying at ±θ degree affected the overall strength in the $[45]_{14}$ case, and as in $[0]_{14}$ there was no difference between the eccentricity and the aligned holes. In $[45]_{14}$, the eccentricity decreased the maximum allowable force by 1.57 percent in the normal eccentricity case and by 15.94 percent in the alternate eccentricity case compared to the aligned case.

Keywords: Carbon-fiber reinforced composites (CFRP), Composites, Two holes, Bending, Joining.

1. Introduction

For decades, composite materials have been widely used in not only the aircraft industry but also in various applications such as automobiles, boats, construction, and in some structural parts of machine components [5-6]. Composites continue to replace conventional materials like steel and aluminum yearly. Engineers have more design flexibility and can steadily improve some strength of their functional parts by using fiber-reinforced composites like fiberglass and carbon fiber. The strength-to-weight ratio is the greatest advantage of composite material usage. It is the key factor for its use in aircraft components because of the weight reduction.

However, the composite also has some disadvantages. First, composites are designed to have superior mechanical properties to those of the constituent material action independently. The properties are unique to each purpose with a specific design for the orientation and manufacturing processes, which provides big challenges to consistently form the structural components. Second, continuous fibers are required for applications at a high performance level. Complex structural sub-assemblies formed by continuous fiber composites must be either attached or assembled with other structural components. There are numerous bonding methods: adhesive, bolts, rivets, welding, etc. From an engineering point of view, adequate bonding strength is usually provided by mechanical fastenings using bolts or rivets. This requires the formation of bolt holes in the drilling process which can cause discontinuity in composite structures [1]. In addition, drill holes have influence stress concentration that locally affects failure initiation, failure mechanisms, failure propagation and the overall behavior of the structural components. This effect can reduce the load carrying capacity of the joint.

The connection concept of this study was used to construct a wing spar for an aerobatic airplane designed in Thailand as a kit plane. The plane manufacturer wanted to build an aircraft where the customer can be involved in the assembly process in creating their own flying machine. Focusing on the wing spar connection, a single hole is not appropriate for a design in terms of strength and reliability. Multiple holes are the only choice for the connection. From the manufacturer point of view, using two holes facilitates the assembly of the airplane. A series of investigative activities in this study was set up [2-4]. However, the research still needs the further experimental results to improve understanding under bending as it is the major load from wing through the connection.

Taking the main loading issues related to the joining pattern, an experimental program was designed to investigate the effects of three different patterns having two equal-sized holes lying at 0, θ, or -θ degrees in a composite plate under bending (Fig.1). The patterns were assigned to be the framework as this is a type of joint that has been intended for use in the construction of the wing spar.

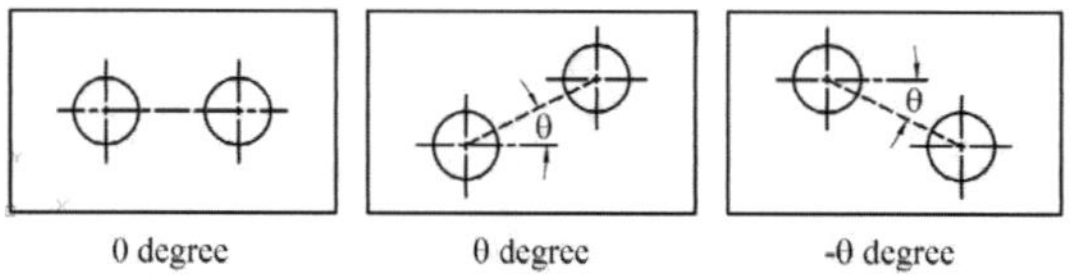

Fig. 1: Hole pattern in two-holed samples

2. Materials and methodology

The selection of materials was defined by the plane manufacturer. The wing spar components were made with carbon-fiber reinforced composites (CFRP). The materials used included a twill 2 x 2 Carbon fiber from Toray carbon fibers America, Inc. and an Epotec YD535LV/TH7257 epoxy resin with low viscosity from Aditya Birla Chemicals (Thailand) Co., Ltd was used as the binding polymer/matrix.

The chosen methods are in this section. At the beginning of the analysis, it is important to realize that the bending strength of two holes has several factors that must be taken into account, including the laminate stacking sequence, laminate thickness, sample size, hole size, hole location, and failure mechanism as well as damage from the fabrication process. It is noted that the failure mechanism is controllable by the laminate stacking sequence and the bearing strength can be more easily identified using a single hole [3]. Therefore, this study used the ASTM D5961 standard test methods to initially define work on the issues of laminate thickness, sample size, hole size, and the failure mechanism. After the bearing strength of a single hole is exhibited, the failure of a two-holed specimen under bending testing can be realized corresponding to the same specimen configurations. Consequently, the experimental procedure covers the following: choosing sample sizes, sample holes, type of testing, hole location, and design jigs and fixtures.

2.1. Description of tested materials

The CFRP plates (250 mm x 430 mm) were fabricated using the resin infusion technique at the laboratory scale. The epoxy has a 3 to 4 hour work life that guarantees fast and complete impregnation of reinforcing fibers and allows laminates to be produced by resin infusion. During the manufacturing processes, the molding was cured at room temperature for 24 hours before it was post cured in an oven at 80°C for 4 hours. Table 1 describes the samples used in this study. For single-holed specimens, the laminate layup configurations were $[0]_{14}$, $[22.5]_{14}$, and $[45]_{14}$. For two-holed specimens, only two different lay-ups of $[0]_{14}$ and $[45]_{14}$ were prepared.

After post curing, the CFRP composite plates were cut to rec-tangular shapes of 36 mm x 210 mm for the single hole tensile test and 45 mm x 210 mm for double-hole bending test using CNC machine. The 3 mm hole was drilled using a CNC machine at the desired position before drilling 5.8 mm and reaming 6.1 mm in diameter. The dimensions of the tensile and bending samples are shown in Fig. 2 to Fig. 4.

Table 1: Test sample description

Test	Ply stacking sequence	No. of holes	No. of specimens	Remarks
Tensile	$[0]_{14}$	1	5	-
	$[22.5]_{14}$	1	5	-
	$[45]_{14}$	1	5	-
Bending	$[0]_{14}$	2	6	Hole 0
	$[0]_{14}$	2	3	Hole θ
	$[0]_{14}$	2	3	Hole -θ
	$[45]_{14}$	2	6	Hole 0
	$[45]_{14}$	2	3	Hole θ
	$[45]_{14}$	2	3	Hole -θ

The averaged volume fraction of the sample was 0.52, which was calculated using Eq.(1):

$$V_f = \frac{\rho_f m_f}{\rho_f (1-m_f) + \rho_r m_r} \qquad (1)$$

where ρ is the density, m is the mass, and the subscripts r and f are the properties of epoxy and carbon fiber, respectively.

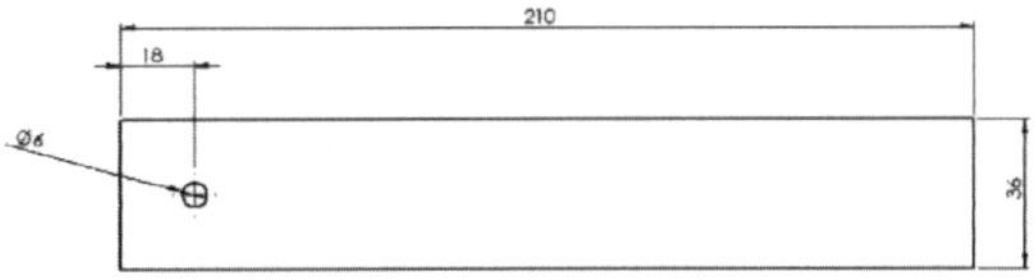

Fig. 2: Dimension of tensile samples

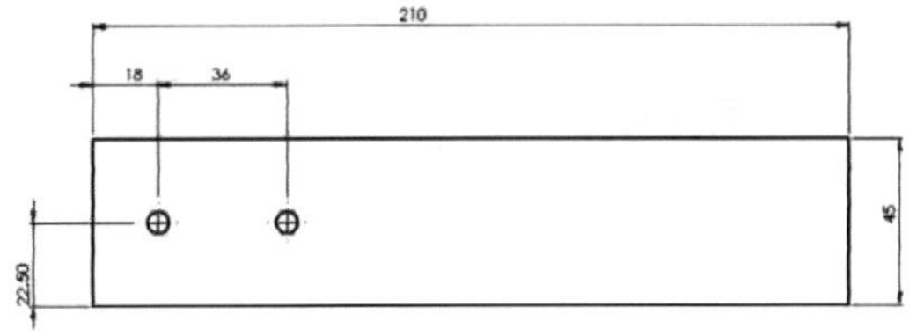

Fig. 3: Dimension of bending samples with two holes lying at 0 degrees

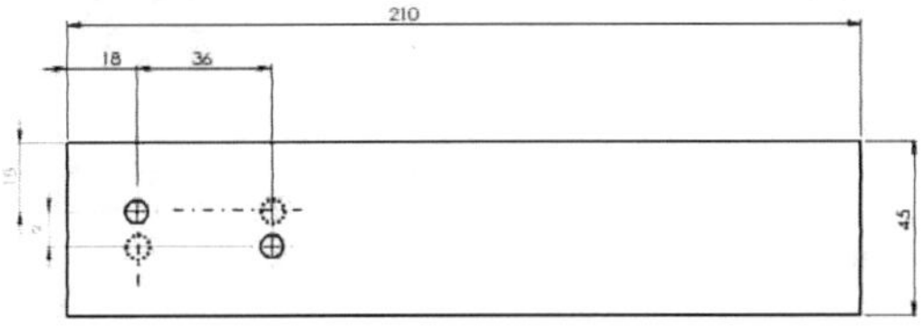

Fig. 4: Dimension of bending samples two holes lying at ±θ degrees

2.2. *Experimental setup*

2.2.1. *Tensile test configuration of single-holed specimens*

In this study, a double shear fixture was designed to apply a load to determine the bearing response of the specimens. One end of the fixture was a plate without a hole while the other end was composed of two rigid plates with a hole. The two rigid plates were set with a spacer slightly thicker than the thickness of the tested specimen. The end of the two plates was drilled with an equal size to one end of the specimens. During the test, a pin was inserted in the three holes by placing the specimen between the two rigid plates of the fixture. The installation of the fixture and a specimen is demonstrated in Fig. 5.

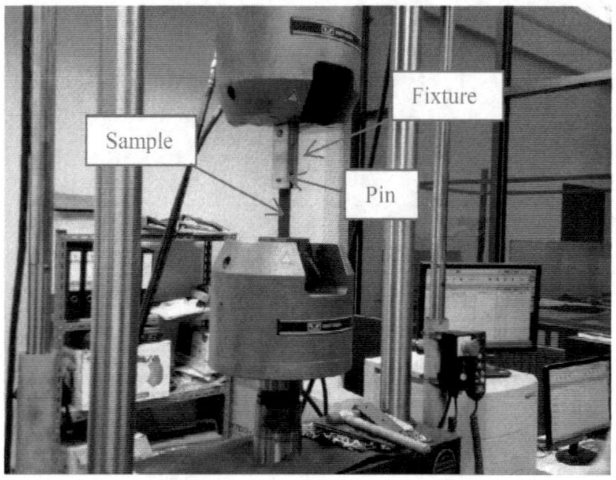

Fig. 5: Test setup for tensile test

Using an Instron 8802 universal testing machine with a maximum load capacity of 250 kN, the bearing force was created by tension loading through the pin. A 6 mm diameter pin was used for the connection. It was installed in the three holes drilled slightly larger than its diameter. One end of the specimen was secured to the bottom vice-grip and the other end of the fixture was held by the upper vice-grip to pull the specimens with a moving speed of 0.5 mm/min. The tensile load was applied until failure was observed.

2.2.2. Bending test configuration of two equal-hole-sized specimens

For the double-holed samples, three-point bending tests were performed on a composite plate specimen using a long arm to generate the bending load through the holes of the sample. The sample was connected with the arm using two 6 mm pins. The distance between the two supporting locations was 1,000 mm. The beam was compressed at mid-span using an MTS/2S universal testing machine with a maximum load capacity of 10 kN. Fig. 6 demonstrates the test set up of three-point bending test for aligning double holes. The moving anvil had a speed of 0.5 mm/min until the sample was broken.

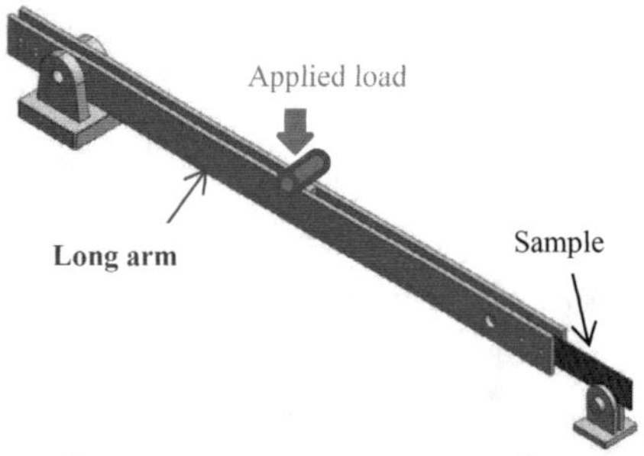

Fig. 6: Test setup for bending test

3. Experimental results and discussion

3.1. Single-holed tensile test

The tensile tests of 15 single-holed samples were carried out for three lay-ups using the fixture in section 2.2.1. The results of each ply stacking sequence and their averages are summarized in Table 2. The averages of maximum force or mean force for the $[0]_{14}$, $[22.5]_{14}$, and $[45]_{14}$ samples were 6.32, 6.59, and 7.12 kN, respectively. The obtained data showed that the $[45]_{14}$ samples had the highest mean force and the $[0]_{14}$ samples had the lowest mean force. The maximum difference between the $[45]_{14}$ and $[0]_{14}$ samples was 11.24%.

Table 2: Single-holed tensile test results

Ply stacking sequence	Maximum force of five samples (kN)	Mean force (kN)
$[0]_{14}$	6.41, 6.60, 6.25, 6.32, 6.02	6.32
$[22.5]_{14}$	6.38, 6.59, 6.53, 6.82, 6.73	6.59
$[45]_{14}$	7.33, 6.53, 6.97, 7.45, 7.31	7.12

The relationships between the load and displacement of the three types of laminate were observed during the test. Between the same ply stacking sequence specimens, the bearing response behavior showed almost the same rigidity and there was a similar magnitude of maximum forces. The bearing response of a specimen representing each lay-up is characterized in Fig.7. It should be noted that the stiffness of each orientation exhibited an opposite trend to the average maximum force. The $[0]_{14}$ samples presented the highest stiffness with the lowest displacement, followed by the $[22.5]_{14}$ samples, and the $[45]_{14}$ samples had the lowest stiffness with the highest displacement. Observing the specimens after the test, it was found that failure would prematurely occur around the hole and propagate along the fiber direction.

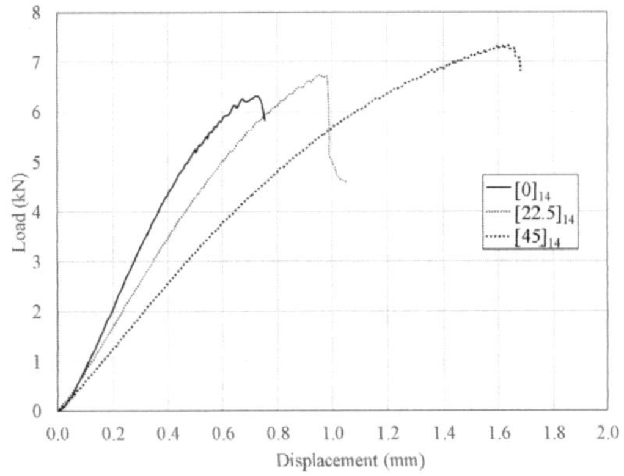

Fig. 7: Load-displacement curve of single-holed specimens

3.2. Result for two equal-sized holes

3.2.1. $[0]_{14}$ cases with hole 0, $+\theta$ and $-\theta$ degree

The relationship of the mid-span force and deflection for the $[0]_{14}$ samples when the holes were drilled at an angle of 0 and $\pm\theta$ degree are presented in Fig. 8 and Fig. 9, respectively. It can be seen in both graphs that the mid-span force linearly increased as the mid-span deflection rises until failure is observed. The maximum mean force for the $[0]_{14}$ laminate configuration was 2.79 kN, for the samples with θ degree hole is 2.89 kN, and for the $-\theta$ degree hole was 2.62 kN. The eccentricity increased the maximum force by 3.51 percent in the $+\theta$ eccentricity case and by a 6.19 percent decrease in the $-\theta$ eccentricity case compared to the aligned case.

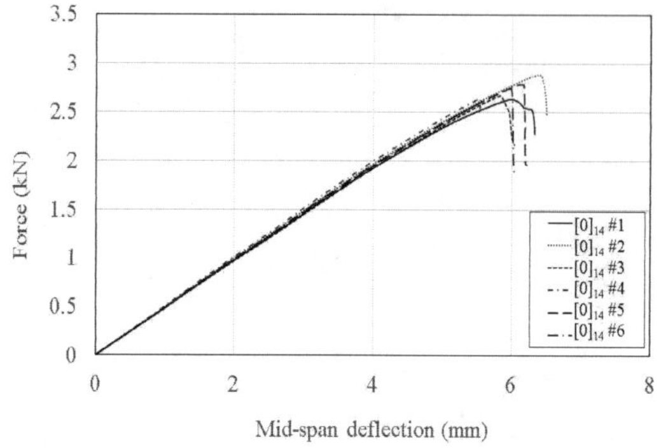

Fig. 8: Mid-span force and deflection relationship for $[0]_{14}$ sample with 0° hole.

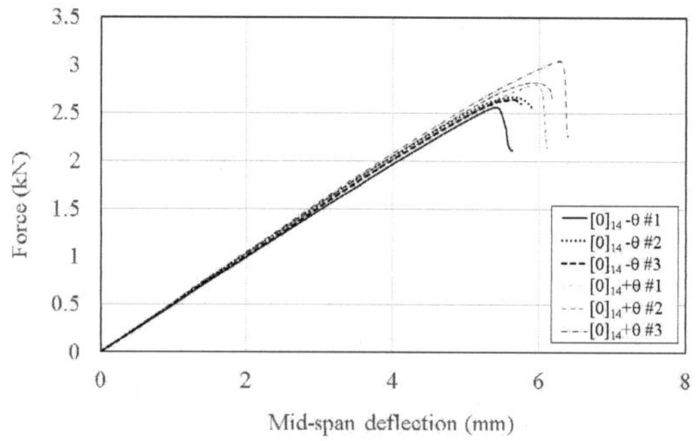

Fig. 9: Mid-span force and deflection relationship for $[0]_{14}$ sample with ±θ° hole.

3.2.2. $[45]_{14}$ cases with hole 0, +θ and -θ degree

The relationship of the mid-span force and deflection for the $[45]_{14}$ samples when the holes were drilled at an angle of 0 and ±θ degree are presented in Fig. 10 and Fig. 11, respectively. A nonlinear response was observed as the bearing responded to the tensile test. The maximum mean forces for 0, +θ, and -θ degrees were 2.32, 2.73, and 2.26 kN, respectively. The eccentricity of the hole at -θ degree increased the maximum mean force by 17.45 percent, while the force was reduced by 2.78 percent in the +θ eccentricity case compared to the aligned case.

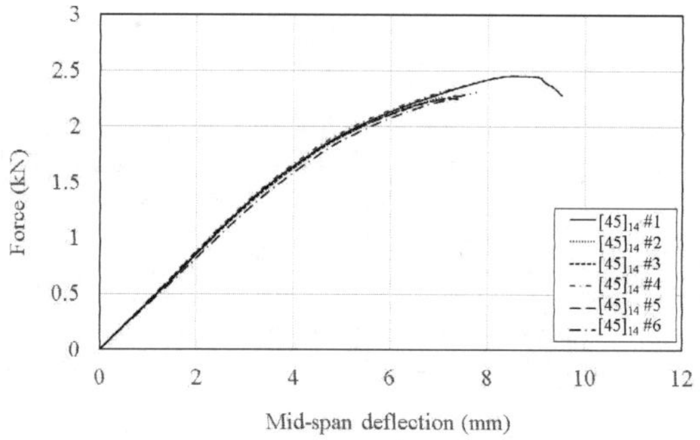

Fig. 10: Mid-span force and deflection relationship for $[45]_{14}$ sample with 0° hole.

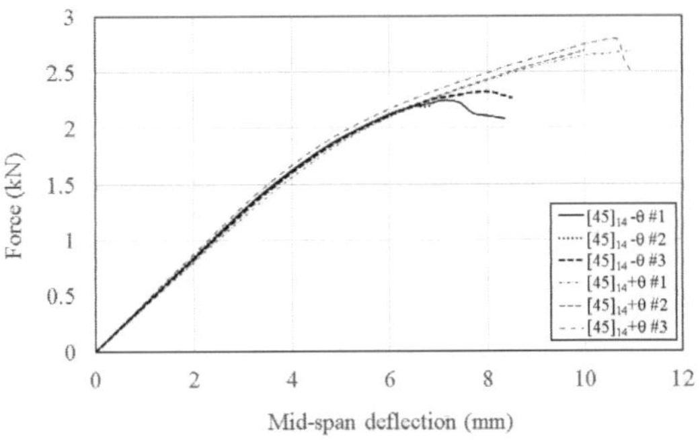

Fig. 11: Mid-span force and deflection relationship for $[45]_{14}$ sample with ±θ° hole.

3.3. Effect of fiber orientation on the bending test

The tensile tests were carried out for $[0]_{14}$, $[22.5]_{14}$, and $[45]_{14}$ ply orientations. Fig.12 presents the failure of the hole under the bearing force. The schematic of carbon-fiber orientation and bearing force are demonstrated. For the $[0]_{14}$ sample, the fiber was oriented in either the loading or transverse directions. Both longitudinal and transverse fibers resisted the bearing load, while the longitudinal one was compressed. For the $[22.5]_{14}$ sample, both the fibers lying along 22.5 and 67.5 degrees compared to the loading direction resisted the bearing load. Both 45 degree orientations of the fiber were advantageous in resisting the bearing force in the $[45]_{14}$ case. This was why the maximum bearing force was presented by the $[45]_{14}$ case and the minimum bearing force was exerted by the $[0]_{14}$ laminate configuration.

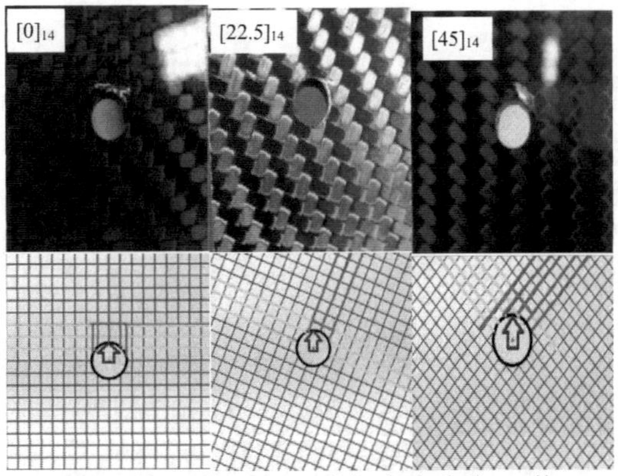

Fig. 12: Failure direction of single-holed tensile test.

Fig. 13 demonstrates the schematic of forces from applying the bending moment to the two holes of the sample in the three-point bending test. It can be seen that there are two reaction forces (R_L and R_R) acting on the overall configuration as shown in Fig. 13(a). Schematics of the forces on the sample are shown in Fig. 13(b) when the load F and moment M are replaced with an equivalent force-couple system at the centroid of the two holes by considering the whole system as a rigid body. It can be found that the applied force on the outer hole, M/d - F/2 must be lower than the applied force on the inner hole, M/d + F/2. Therefore, the inner hole should initially fail due to bearing force before that occurs at the outer one. It should be noted that the mass of the samples is diminished in the schematic because it is too small when compared to the mass of the long arm.

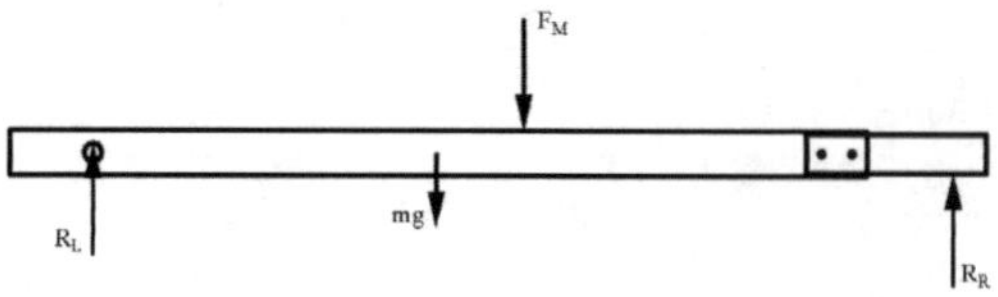

(a) Schematic of force on the bending test

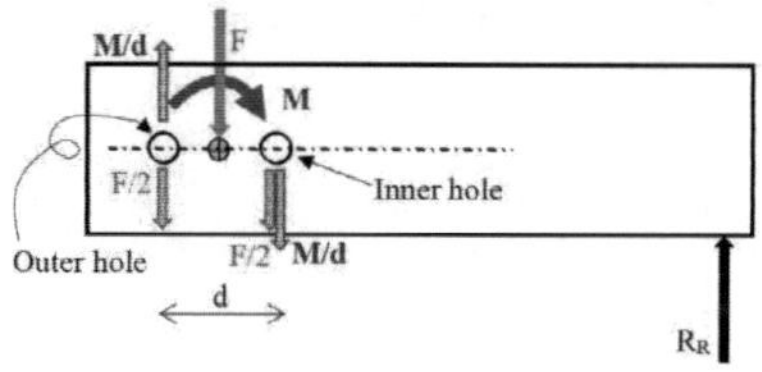

(b) Schematic of forces on the sample

Fig. 13: Bending test configuration

The bending tests were carried out for $[0]_{14}$ and $[45]_{14}$ ply orientations. Three different patterns were tested having two equal-sized holes lying at 0, θ, or -θ degrees in a composite plate as shown in Fig.1. It was found that the failure occurred in the samples as the inner hole rotated about the outer hole in every case. Fig.14 presents the orientation of the force along the inner hole for the three degree cases by applying equilibrium analyses on the resulting free-bodies to determine the internal shear forces and bending couples. It can be seen that the bearing load was exerted on the inner hole with angles of 90°, 12.7°, and 12.7° for the samples having the two holes lying at 0°, θ°, and -θ°, respectively. The inner hole must carry both the force and moment from the shear force and bending moment developed within the beam. The -θ degree samples had the lowest strength under bending in the tests. With the limitation on edge distance, there is a higher failure potential for the -θ degree samples than the other proposed patterns of the hole.

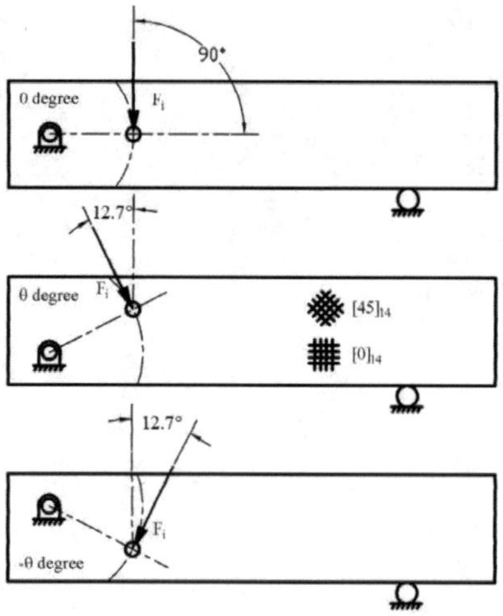

Fig. 14: Orientation of bearing force on inner hole for bending test.

4. Conclusions

The present study investigated the bearing strength of single-holed specimens in accordance with the ASTM D5961 standard test. From the laboratory testing, it was possible to conclude that crack formation is likely due to the orientation of the fiber. For the $[0]_{14}$ samples, where the load was applied along the fiber direction, the failures occurred at the contact area and then propagated in the load direction because of the load carried by the fibers. For the $[22.5]_{14}$ and $[45]_{14}$ samples, the direction of the fibers was inclined to the load direction. The failure presented at the resin supporting the bearing load at the contact area which then propagated along the fiber direction which was inclined to the applying load. The test results indicated that the $[0]_{14}$ sample had the highest maximum bearing force with the lowest displacement. This implied that CFRP is brittle in nature and thus prone to cracking along the fiber direction.

The above-mentioned investigations provide a general guideline for joints where a single hole is applied to the structure. The concept of the failure of the single-holed samples under tension is considered in the same manner as the two-holed sample under bending. A comparison of the failure initiation for the three different hole patterns with two lay-ups was observed. The force-deflection curve showed non-linear behavior for the $[45]_{14}$ samples. The samples with two holes lying at -θ degrees had slightly more resistance under bending than the other hole patterns. The results revealed that failure corresponded to the orientation of the fiber.

Acknowledgement
This research was financially supported in part by the Kasetsart University Research and Development Institute (KURDI), Bangkok, Thailand (document number 20420186110223000). The authors would like to express their sincere gratitude to KURDI and to everyone who made a contribution to this study.

References

[1] Durao L, Tavares J, De Albuquerque V, Marques J & Andreade O. Drilling Damage in Composite Material. Materials (ISSN 1996-1944), Vol.7, No.5, 2014; pp.3802-3819. available online: https://www.ncbi.nlm.nih.gov/pmc/articles/PMC5453231/, last visit:08.02.2019.
[2] Krajangsawasdi N, & Bunyawanichakul P. Analysis on optimal location of double-bolted joint with parallel misalignment Subjected to eccentric load. Proceedings of 55th Kasetsart University Annual Conference: Science and Genetic Engineering, Architecture and Engineering, Agro-Industry, Natural Resources and Environment 2017; pp:432-442. ISBN: 978-616-278-368-5.
[3] Krajangsawasdi N, & Bunyawanichakul P. Experimental study on angle orientation to the failure of GFRP composite. The 2nd Southeast Asia Workshop on Aerospace Engineering SAWAE 2017.
[4] Krajangsawasdi N, & Bunyawanichakul P. Strength of composite plate with parallel offset misalignment double-bolted joint under bending moment. MATEC Web Conference issued on the 2nd International Conference on Composite Material, Polymer Science and Engineering CMPSE2018; Vol.264, No.01002. doi: 10.1051/matecconf/201926401002.
[5] Nedelcu R, & Redon P. Composites materials for aviation industry. The 14th International Conference Scientific Research and Education in the Air Force-AFASES 2012; 917–922.
[6] Tsai S. Theory of composites design 3rd edition. 2008; pp:_A1-7.

Study and Evaluation of a Solar Floating Traction under Severe Wind Conditions

Pongwit Siribodhi*, Phacharaporn Bunyawanichakul

Design Clinic Research Unit, Department of Aerospace Engineering, Faculty of Engineering, Kasetsart University, 50 Ngam Wong Wan Rd., Ladyaow, Chatuchak, Bangkok, 10900, Thailand.

*Corresponding author e-mail:fengpws@ku.ac.th

Abstract. Solar cells have been among the first such clean energy groups that have been used, talked about, and adopted extensively in many areas. One drawback of using solar cells for energy supply not only in households but also in large cities is the space needed as well as usually requiring a flat area. For these reasons, the construction of solar farms over water is a popular concept and continues to be of interest in Thailand and the rest of the world. This research provided a case study of implementing large-scale solar cells in Southeast Asia where unexpected typhoons are common during the monsoon season. This can be addressed by binding the solar panels together to make a large platform that can remain intact in the water under various conditions. Engineers need to be confident that such binding will be safe enough to ensure the solar floating platform has no movement which may cause unexpected damage. This study developed and evaluated solar panel traction with an arrangement of 9 x 28 and 28 x 9 panels under severe wind conditions of 120 kilometers per hour (33.33 meters per second) which is equivalent to typhoons in the region. The evaluation used empirical and numerical models, and wind-tunnel testing. The resultant equation was: $D(N, M) = 2075.8\ N^{0.6532}\ M^{0.9052}$ and $D = Kv(V)\ 2075.8\ N^{0.6532}\ M^{0.9052}$ where N and M are the numbers of platforms in columns and rows, respectively, $Kv(V) = (0.019564618+0.007704667V)/(1- 0.0004656279V)$, and V is the wind speed (km/h). The investigation found that the predicted traction results using the derived equations was in good agreement with wind-tunnel testing for a proposed platform.

Keywords: Floating solar farm, Mooring system, Solar cell, Floating platform, Solar floating aerodynamics.

1. Introduction

There has been a steady increase in human energy consumption around the world. Previous generations had lower energy requirements than humans at present. It is clear that almost everyone has personal gadgets and smart appliances even with a modest lifestyle, which is leading to higher and higher everyday energy consumption. In 2015, a review reported that humans generally consume as much as 110 times as much energy per person as our early ancestors [1]. Nowadays, the main source of energy come from fossil fuels (coal, natural gas, oil). However, fossil fuels directly impact on the environment and they are nonrenewable resources. Therefore, a major step forward for humans, as the world's biggest polluter, is to in-vest in renewable energy sources. Clean energy is considered to be an alternative to traditional power sources such as coal, natural gas, and oil around the world.

There are many sources of renewable energy including solar energy, wind energy, hydropower, and bioenergy. Among these, solar and wind energy are exceedingly cheap to operate and their sources of energy are free. According to the World Energy Council [2], "the total energy from solar radiation falling on the earth was more than 7,500 times the World's total annual primary energy consumption of 450 EJ" [3]. For this reason why people are looking for new developments to convert sunlight directly into electricity.

The installation of solar panels is more complex and difficult on a water surface requiring a different system than on land. A distinct difference is the aerodynamic loading and various unpredictable damage patterns. In addition, solar panel installation is generally used to produce large-scale capacity and requires an appropriate large, open space for ensuring energy production. A study concluded that a water surface installed with solar cells was 11% more efficient in electricity production than onshore solar cells [4]. The challenge in constructing solar panels on the water surface is related to the design, structural strength, and binding of the platform. In Southeast Asia every year, there are typhoons during the monsoon season. An average typhoon wind speed is approximately 120 kilometers per hour (33.33 meters per second) [5]. This means that a solar cell platform will be subject to aerodynamic loads from such wind speeds which can result in a traction force on the platform. Such a traction force, if high enough, may cause the platform to move or fall out of the alignment and lead to damage to the solar cells from collisions.

2. Scope of work

Since solar panels only generate electricity when the sun shines, to produce enough energy it is necessary to have a large area of solar panels apart from considering the efficiency of the solar cells in the panels. The more solar cells working, the more power they create. However, most manufacturers fabricate solar panels at a marketable size. It is essential to arrange a group of solar panels on a huge surface for the purpose of efficiency. In this study, 1,512 sets of solar cells were installed on a platform floating on a water surface. Each set was composed of 12 solar panels disregarding the under-water part as calculated using the static equilibrium equation [1,6-7] as shown in Fig.1. Different patterns in terms of the numbers of platforms in columns and rows, represented by N and M, respectively, were established to fit with the reservoir surface. An example of 15 sets where N x M equals 5 x 3 is illustrated in Fig.2.

Fig. 1: Solar floating platform set

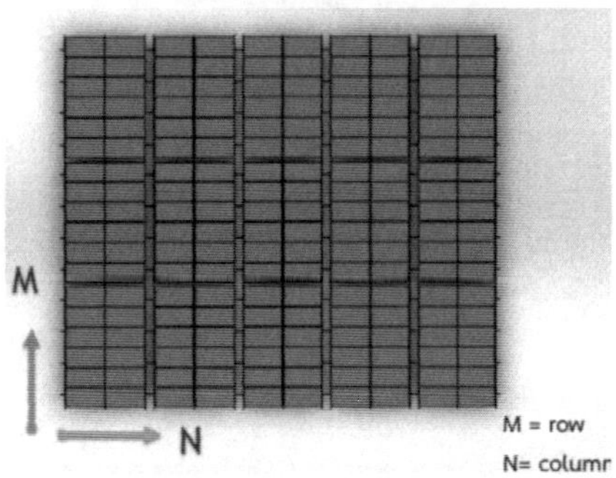

Fig. 2: Arrangement of N x M representing 5 x 3

In this paper, the solar cell platform was actually constructed with several groups of 252 sets established in two main patterns of 9 x 28 and 28 x 9. The traction that occurred on the platform under the aerodynamic load at a wind speed of 120 kilometers per hour was calculated and evaluated. The investigation was performed on a 1:20 scaled model using wind tunnel testing in order to validate the calculation results. Finally, a closed form traction force equation with independent variables consisting of the number of platforms in a row, the number of platforms in a column and the wind speed was established. This equation was used to develop traction force criteria for estimating the anchor strength of the platform arranged in an operating pattern of 9 x 28 and 28 x 9.

3. Methodology

This study used three tools to calculate and evaluate the traction force on the platform: a closed form equation, computational fluid dynamics, and a wind tunnel test. There are recognized limitations in assessing traction force using each technique. Therefore, this study took advantage of the best components of each tool and these were used to validate each other. The results of the study can be summarized as follows:

3.1. Closed form equations [8]

According to the American Society of Civil Engineering Standard issued on ASCE7-10 Minimum Design Loads for Buildings and Other Structures, an open building with gable roof is subject to wind load as defined in Fig.3. The wind direction come from the left with no angle to the horizontal level. Two net pressure coefficients, C_{NW} and C_{NL}, contributed from the top and bottom surfaces, shall be considered when the wind speed comes from the side of the buildings. The building has a horizontal dimension of roof, L, measured in the along wind direction, mean roof height, h, angle of plane of roof from horizontal, θ and wind direction, $\gamma (= 0°)$. For the case of an open building, the velocity pressure, q_h, evaluated at the mean roof height h was calculated in SI unit using the following equation:

$$q_h = 0.613 K_h K_{ht} K_d V^2 \quad (1)$$

where
K_h = velocity pressure coefficient = 1.03 (exposure d-flat, unobstructed areas and water surfaces at height 0-4.6 m)
K_{ht} = geographic factors = 1
K_d = wind directionality factor = 0.85 (main wind force resisting system; MWFRS)
V = wind speed = 33.33 m/s

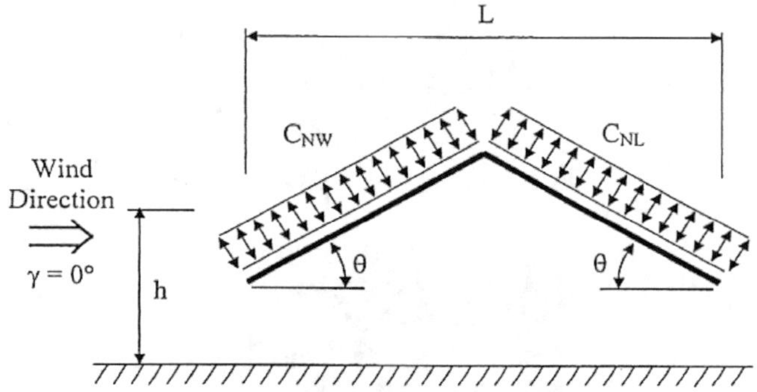

Fig. 3: Wind loads on open buildings

For open buildings, the design pressure, p, was determined from the following equation:

$$p = q_h G C_N \quad (2)$$

where
q_h = velocity pressure evaluated at mean roof height h, determined from Eq. (1)
G = gust-effect factor = 0.85
C_N = net pressure coefficient, for angle of plane of roof from horizontal, $\theta = 12°$, the values of C_{NW} = 1.1 and C_{NL} = -0.36

Two 12-degree slopes of the gable roof ($\theta = 12°$) for this case and the effective solar panel area of 12.59 m^2 were considered to determine the total traction acting on the structure.

3.2. *Computational fluid dynamics*
Computational fluid dynamics (CFD) is another engineering tool that can be used for the prediction of the traction from the aerodynamic characteristics of the platform in this section. The typical size for a set was 4 meters by 6 meters and the depth was 0.77 meters (Fig.4). The computational domain and the model disregarding the sunken part are presented as illustrated in Fig.5.

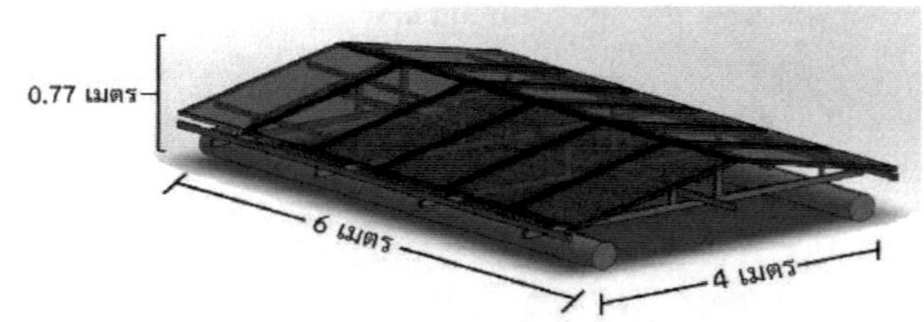

Fig. 4: Dimension of a solar floating platform set

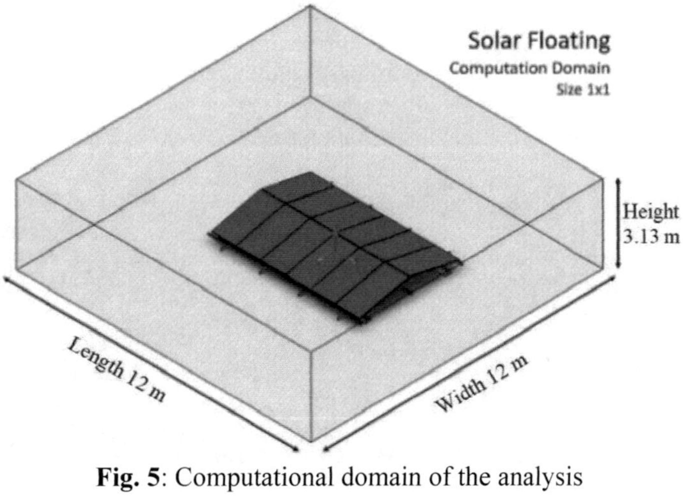

Fig. 5: Computational domain of the analysis

3.3. *Wind-tunnel testing*
A conceptual and preliminary study was undertaken concerning the experimental design before fabricating the models and other components. The size of the test section, wind speed, depth of water, ramp angle for smooth flow, and load cell specification are affected by the characteristic of the solar floating system under wind conditions and diagnostic instrumentation. CFD analysis played an important role in this early stage. Fig.6 shows the dimensions of the designed water tray that was fixed at 2.06 m x 0.81 m to fit with the test section of the tunnel (Fig.6). These dimensions also included a ramp for making the flow as smooth as possible.

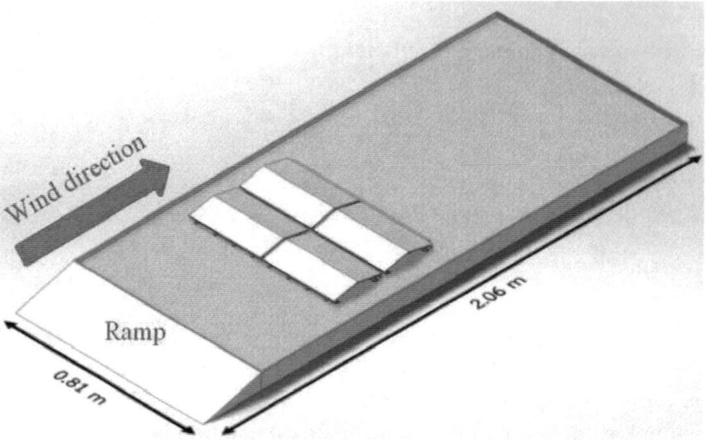

Fig. 6: Various components placed in the wind tunnel

The model and other components for wind tunnel testing studied were: ramp slope, wind speed, water depth and, especially the arrangement of the solar floating platform. All parameters had to allow for the load cell capability and for the traction to be measured. The study found that the 1:20 scaled platform was suitable for testing in a wind tunnel. A water tray including a ramp with an angle of 15 degrees was appropriate to provide a smooth flow. A single set of the solar floating platform (the 1 x 1 model) resulted in too little traction to be measured by the existing load cell. Therefore, a group of scaled models with an arrangement of 2 x 2 was used to increase the measured drag force approximated by the CFD analysis within the load cell limitation (Fig.7). Moreover, the analytical closed form equation could not be used to determine the drag force on the 2 x 2 model and so the drag forces were measured directly using a load cell connected to a special adaptor which was designed for this test purpose as shown in Fig.7. The preliminary study using CFD indicated that the wind direction approaching the platform could have a significant influence on traction. Consequently, wind speeds of 5, 7, 9, and 11 m/s were used for the test using both a side (Fig.8) and front (Fig.9) wind direction.

Fig. 7: Various components placed in the tunnel during the test

IOP Conf. Series: Materials Science and Engineering **642** (2019) 012003 doi:10.1088/1757-899X/642/1/012003

Fig. 8: Testing the 2 x 2 model with side wind direction

Fig. 9: Testing the 2 x 2 model with front wind direction

4. Results and discussions

The normal pressure on the gable roof was calculated using eq. (1) and (2). The calculation was performed for an open structure representing a single set of the solar floating platform or the 1 x 1 arrangement according to ASCE7-10. It was found that normal pressure exhibited in Fig.10 was not equal to the windward and leeward halves of the roof surface. When the wind came from the left side, the windward roof facing the wind first had higher normal pressure than the other roof. According to eq. (2), the normal pressure acting on the left and right roofs was 557 N/m2 and 203 N/m2, respectively, indicating that the resultant traction from the closed form equations of a single platform under 120 km/h (33.33 m/s) was 1,936.71 N.

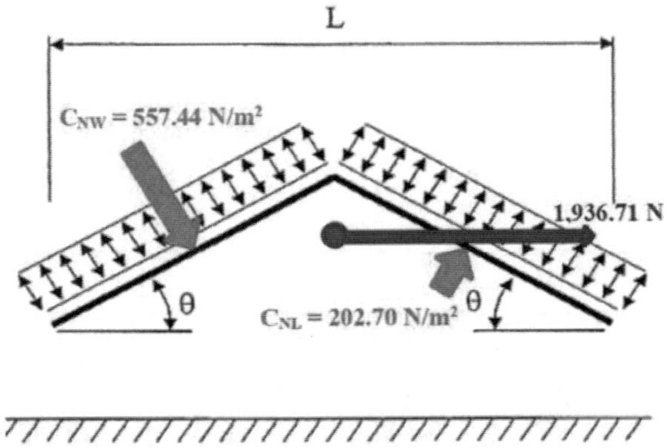

Fig. 10: Pressure acting on gable roof of an open building resulting in traction on the 1 x 1 platform

Using CFD, the mathematical model of a single platform, the 1 x 1 model without sunken parts, was analyzed with an inlet velocity of 120 km/h (33.33 m/s) which was the same condition used in the closed form analysis. The mesh of the model was generated as rep-resented in Fig.11. Simulation was used to extract the drag force exerted on the model. The total drag force acting on the entire sur-face area of the model by the resultant force in the flow direction was 2,153.27 N.

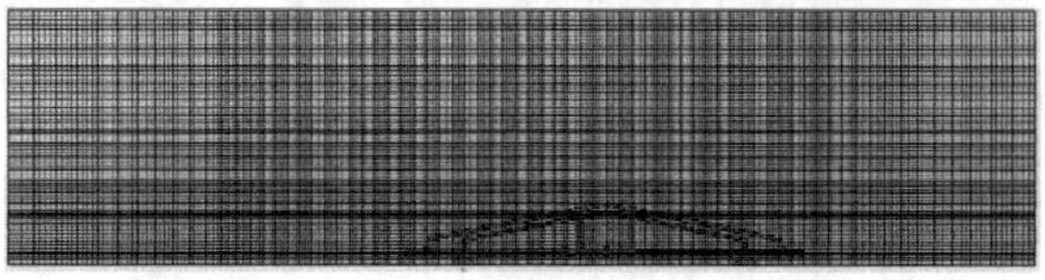

Fig. 11: Mesh of the 1 x 1 model

Based on the CFD simulation results, the speed in a given wind direction of the air throughout the model is demonstrated in Fig.12. It indicates that the windward roof directly confronts the flow while the flow is separated on the leeward roof. This study confirmed a similar trend to the traction obtained from the closed form analysis.

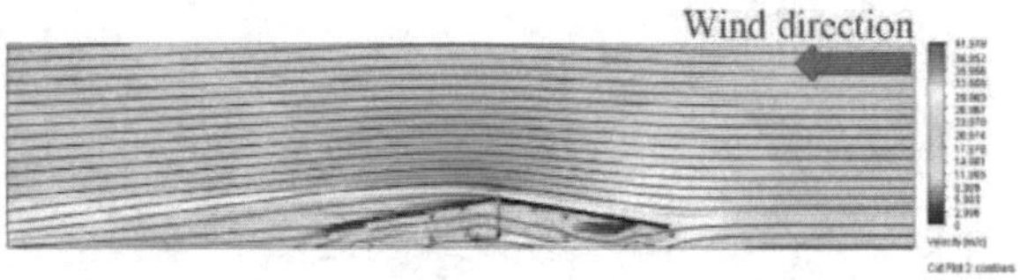

Fig. 12: Velocity display within the computational domain

The results of the 3 evaluation methods for the case of the 1x1 plat-form for both the closed form analytical equation and the CFD are shown in Table 1. The difference between the 2 cases was approximately 10%. The traction obtained from the standard was less than from the CFD analysis probably due to the standard taking only the roof into account while the CFD analysis considered the whole floating part in the mathematical model. These results show the prediction of the traction was within the range of 1,900 N to 2,200 N which is a good beginning for drawing a set of solar floating platforms on a water surface. The results indicated that the traction was greater in the CFD result than the standard. It is preferable to rely on the value from CFD analysis.

Table 1: Difference between traction obtained from the ASCE 7-10 standard and CFD analysis for a solar floating platform set

ASCE 7-10	CFD	% Difference
1,936.71 N	*2,153.27* N	10.06

The traction was estimated using CFD analysis of a mathematical model developed during previous study. The model was simulated only for a single unit of the solar floating platform. In this case, a group of 2 x 2 pattern had been tested in the wind tunnel to compare the traction obtained from CFD analysis.

This work developed a mathematical model, including the floating part of the 2 x 2 pattern of a 1:20 scaled model, a water tray with a ramp containing water of a fixed depth. The sunken part of the scaled model was calculated using static equilibrium. Two wind directions were studied: side and front. The simulation was run with air speeds of 5, 7, and 9 m/s. There were problems associated with using the 1:20 scaled model, including difficulty in making the model float in a realistic manner and because of the scaling, with-standing wind speeds greater than 9 m/s. Following the failure due to diverging oscillation in the wind tunnel caused by the instability, it was only possible to test using 5, 7, and 9 m/s for comparison with the CFD simulation. The experiment was conducted in the low speed wind tunnel at the Department of Aerospace Engineering, Kasetsart University, Bangkok, Thailand.

Flow trajectories over the model with an air speed of 7 m/s to the side direction are shown in Fig.13. It can be seen that the velocity increased as the air collided with the ramp. Subsequently, the velocity gradually increased to the top of the ramp. At the same time, the velocity directly behind the ramp dropped substantially to ap-proximately 4 m/s. This velocity accelerated to 9 m/s when the air flowed over the windward roof of the 2 x 2 scaled platform. The air speed substantially decreased over the leeward roof. After this point, the air speed was forced to increase over the next gable roof, whose inclined slope was the same as the windward roof. However, the air speed at the top of the second gable was lower than for the first. This indicates that the kinetic energy of the wind can be dissipated when the air stream faced the first gable. For groups of several sets of solar floating platforms, it is possible to reduce wind resistance by installation of solar floating sets close behind each other.

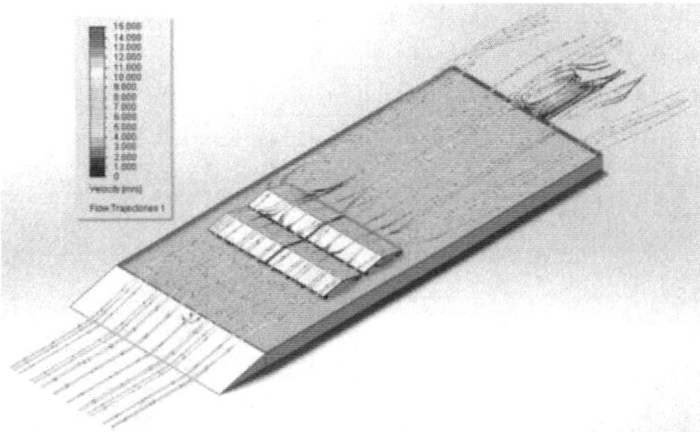

Fig. 13: Flow trajectories of the 2 x 2 pattern of 1:20 scaled model

The averaged results of the CFD simulations versus the wind-tunnel measurement are presented in Table 2 for the 2 x 2 pattern of the 1:20 scaled model. The side wind direction had more influence on the traction than the front wind direction. It was clear that these results yielded from the frontal area of the model in both cases. The trend line can be obtained by using the quadratic equation shown in Fig.14.

Table 2: Comparison of traction obtained from CFD simulations versus wind-tunnel measurement for the 2 x 2 pattern of 1:20 scaled model

Velocity (m/s)	Traction (N)			
	Side wind direction		Front wind direction (N)	
	CFD-Side	WT-Side	CFD-Front	WT-Front
5	0.14	0.17	0.10	0.12
7	0.26	0.29	0.15	0.27
9	0.44	0.46	0.25	0.38

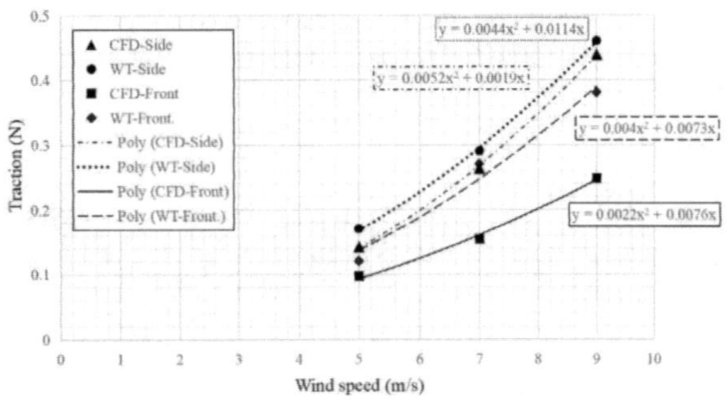

Fig. 14: Comparison of traction results from side and front wind directions

For a side wind direction, there was good agreement between the experimental and simulation results. For a front wind direction, the results indicated that increasing the wind speed could generate greater traction as was observed during the tests carried out. It was noted that the scaled platform could easily vibrate on the water due to wind and waves. When the wind speed increased, the waves oscillated up and down with a higher amplitude. This motion induced instability in the whole platform, especially, forming an angle of attack which had a large effect on the traction.

5. Conclusion

The agreement between the CFD analysis and the wind tunnel measurements was considered good, which could support using these simulations for further analysis of the flow field and also using the same computational models for the CFD analysis of groups of on-site platforms. In this case, a group of 9 x 28 and 28 x 9 pattern was used on site. It was not possible to generate a mathematical model of the 252 units in the computational domain due to the time required.

The study results showed that the CFD model used was confirmed by both the close form equation and the experimental model. By using the CFD model, the research was expanded into a full scaled analysis of an N x M arrangement and could summarize the traction equation as a function of the pattern of the platform as follows:

$$D\,(N, M) = 2075.8\, N^{0.6532}\, M^{0.9052} \quad (3)$$

for a wind speed of 120 km/hr (33.33 m/s)

and

$$D\,(N, M, V) = K_v\,(V)\; 2075.8\, N^{0.6532}\, M^{0.9052} \quad (4)$$

where
$K_v\,(V) = (0.019564618+0.007704667V) \,/\, (1\text{-}0.0004656279V)$
D = drag or traction force (newton)
N = number of platforms in a column
M = number of platforms in a row
V = wind speed (km/h)

Further study is needed on the vibration effect on several sets of the solar floating platform. In addition, traction might be reduced by the installation of each solar floating set at a separation distance using the CFD analysis.

Acknowledgement
The authors thank Kampot Cement Co., Ltd. for supporting this research and making the learning experience possible through the grant reference no.7010161755 to the Design Clinic Research Unit (DSCN), Department of Aerospace Engineering, Faculty of Engineering, Kasetsart University, Bangkok, Thailand.

References

[1] Nelson S. Physical Geology, Energy Resources: Tulane University 2015. available online: https://www.tulane.edu/~sanelson/eens1110/energy.htm, last visit:08.02.2019.
[2] World Energy Council. World Energy Resources: Hydro 2013. available online: https://www.worldenergy.org/wp-content/uploads/2013/10/WER_2013_5_Hydro.pdf, last visit:08.02.2019.
[3] Urban F & Mitchell T. Climate change, disasters and electricity generation. Institute of Development Studies IDS 2011.
[4] Choi Y-K. A Study on Power Generation Analysis of Floating PV System. International Journal

of Software Engineering and its Applications 2014; Vol.8 (1):75-84. doi: 10.14257/ijseia.2014.8.1.07

[5] World Meteorological Organization Technical Document. Typhoon Committee Operational Manual: Meteorological Component 2015. Tropical Cyclon Programme, Report No. TCP-23.

[6] Jompuk K, Triyakun N, Bunyawanichakul P & Siribodhi P. Drag Analysis of Solar Floating under Aerodynamic Load. Proceedings of the 32nd Conference of Mechanical Engineering Network of Thailand 2018.

[7] Pornvoraphat P, Triyakum N, Bunyawanichakul P & Siribodhi P. Lift Analysis of Solar Floating under Aerodynamic Load. Proceedings of the 32nd Conference of Mechanical Engineering Network of Thailand 2018.

[8] American Society of Civil Engineers. Minimum Design Loads for Buildings and Other Structures. Structural Engineering Institute 2010:250-262.

Application of Structural Similitude for Scaling of a Pressure Vessel

Beenish Batul[1], Abeera Sohail[2], Ahmad Aizaz[3], Zukhraf Jamil[4]

[1, 3, 4,] Institute of Aviation Studies, University of Management and Technology, Lahore, Pakistan.
[2] Institute of Space Technology, Islamabad, Pakistan.

*Corresponding author e-mail: beenish.batul@umt.edu.pk

Abstract. Pressure vessels find their use in various fields, ranging from gas cylinders used in households for cooking, boilers for steam engines, fuselage of aircraft to solid rocket motors used in missiles and space shuttle. The design of such vessels is validated by performing tests on full scale prototypes. Mostly the testing of such vessels is cumbersome and expensive. This paper establishes the method to reduce the cost for testing the pressure vessels. The theory of similitude is studied to make the testing process easier by establishing structural similitude for a pressure vessel. Using similitude theory a scaled model of the prototype vessel is developed in such a way that when the scaled models' responses are multiplied by a calculated scale factor, behaviour of the prototype could be predicted. By testing on the scaled down model, the cost of manufacturing is reduced. The pressure vessel considered here is representative of the pressures and materials used in high pressure applications. In this paper a 1/10th scaled model of the pressure vessel is developed using structural similitude theory. Buckingham pi-theorem technique has been used for dimensional analysis after studying parameters on which pressure vessel is designed and ANSYS software is used to validate the resulting pi-products. Complete similarity is achieved when predicted prototype results completely map on to prototype results.

Keywords: Dimensional Analysis, Pressure Vessels, Scaled Model, Structural Similitude.

1. Introduction

In all modern manufacturing processes extensive testing and experimental verification is needed to ensure reliability and safety of the finished structure. Huge constructions like dams, bridges, and transport aircrafts require full scale testing of large components called prototypes to verify theoretical predictions about systems' behavior under various operating conditions before they are considered acceptable. This testing can be expensive and time consuming and remains an indispensable part of modern engineering. If accurate predictions about prototype behavior can be made using experimental results of scaled down easier to handle models, valuable time and money can be saved. This is the motivation behind similitude of models and this study.

A model of a prototype is a system which replicates the behavior of the prototype to a specific input. The experimental results of the model may then be extrapolated to predict the experimental response of the prototype using a relationship which exists between the parameters and variables of the model and prototype called scale factors. The accuracy of the results depends upon the material, design and analysis techniques of the model. These predicted experimental results are then compared with theoretical prototype results.

The similitude of systems which predicts the structural behavior of large components by use of models is termed as structural similitude [1]. The structural similitude theory is already being applied

IOP Conf. Series: Materials Science and Engineering **642** (2019) 012004 doi:10.1088/1757-899X/642/1/012004

in structural [2], vibration, and impact [3] problems in civil and mechanical engineering. It has also been investigated for developing scaled models of stiffened cylinders and shells for various applications and operating conditions [4-5]. In this paper, the structural similitude is developed for thin walled pressure vessels used in aerospace applications and the similarity obtained is studied.

2. Similitude Theory

A mathematical model of a system can be made using all the system variables and parameters. For similar systems these parameters are similar but not identical. The theory of similitude searches for a relationship (transformation) which maps the models parameters onto the prototypes parameters. This means that each model parameter is proportional to its corresponding prototype parameter. This proportional factor is called the scale factor. For example if vectors X_{pr} and X_m are the characteristic vectors of the prototype and model, then we can find a transformation matrix Λ [6] such that:

$$X_{pr} = \Lambda X_m \text{ or } X_m = \Lambda^{-1} X_{pr} \tag{1}$$

In its simplest form diagonal elements of Λ are the scale factors. This need not be the case always.

$$\lambda_{xi} = \frac{x_{ipr}}{x_{im}} \tag{2}$$

$$\Lambda = \begin{bmatrix} \lambda_{x1} & 0 & \cdots & 0 \\ 0 & \lambda_{x2} & \cdots & 0 \\ \vdots & \vdots & \ddots & \vdots \\ 0 & 0 & \cdots & \lambda_{xn} \end{bmatrix} \tag{3}$$

There are two methods which are used to determine scale factors for scale model development of cylindrical shells and other structures: direct use of governing equations and dimensional analysis [7]. When the number of design parameters is large, dimensional analysis can be tedious so instead the systems' governing equations along with the appropriate boundary and initial conditions are used to find the similarity conditions [8]. If the governing equations of the two systems can be mapped onto each other (are proportional) then the two systems are similar and the response of one can be used to find that of the other. Dimensional analysis method is used either when the mathematical model of the system is unavailable or the numbers of parameters involved are few. It requires that all parameters which affect the systems' response to a particular input be known. The similarity conditions are in the form of non-dimensional constants which remain same for both prototype and model and are used to decide parameters and test conditions of the model. The technique used most frequently to find these constants is called the Buckingham pi theorem [9] which is used in this paper.

3. Pressure Vessels

The American society for mechanical engineers defines pressure vessels as a containment of solid, liquid or gaseous material under internal or external pressure, also capable of withstanding various other loadings [10]. Some examples of pressure vessels use are in aircrafts, solid rocket motors, nuclear reactor vessel, diving cylinder, distillation towers and many vessels in mining or oil refineries and petrochemical plants.

Pressure vessels carry liquids and gases at high pressure and so withstand pressure loading which creates stresses and strains. For most engineering applications the thin wall pressure vessel can be used which by definition has a ratio of r/t ≥10.

For this study, a thin wall pressure vessel loaded internally is considered. It is capped with ellipsoidal heads which are made of the same material as the cylindrical portion and is of same thickness. The pressure considered is 5 MPa which makes it suitable for high pressure applications. The mean radius of the cylinder (and the caps) is 'R'. The cylinder and heads have a uniform thickness and the vessel is subjected to an internal pressure and a zero external pressure. No other external forces act. The vessel walls are constructed of a single material.

3.1. Stress equation for thin walled pressure vessels
In a region of the cylinder that is far from the ends, three normal stresses (σ_h, σ_l and σ_r) may be calculated to characterize the thin shell stress state [11]. Therefore:

$$\sigma_h = \frac{pR}{t} \tag{4}$$

$$\sigma_l = \frac{pR}{2t} \tag{5}$$

$$\sigma_r = -(\frac{1}{2}\, p) \tag{6}$$

3.2. Strain equation in cylinder
Total radial displacement is caused by a combination of all three stresses and called 'Δr' here [11]:

$$\Delta r = \left(\frac{pR^2}{2\,Et}\right)\left[2 - \nu + \nu\left(\frac{t}{R}\right)\right] \tag{7}$$

The displacement of the cylinder in the longitudinal direction is given by [12]:

$$\Delta z = \frac{pRy}{Et}\,(0.5 - \nu) \tag{8}$$

3.3. Normal stresses for ellipsoidal head
Longitudinal stress in the head is:

$$\sigma_l = \frac{pa}{2t} \tag{9}$$

Since semi major axis of head will be equal to cylinder radius R, ellipsoidal longitudinal stress is same as longitudinal stress of cylinder. Hoops stress is given by:

$$\sigma_h = \frac{pa(2b^2 - a^2)}{2tb^2} \tag{10}$$

3.4. Strain in ellipsoidal head
The longitudinal strain in the head is small when compared to the cylinder and is hence not considered. Radial displacement for ellipsoidal head at the equator is given by:

$$\Delta r = \frac{pR^2}{Et}\left(1 - \frac{a^2}{2b^2} - \frac{\nu}{2}\right) \tag{11}$$

It is important to note that the net displacement of the cylinder radially will be a value between radial displacement for cylinder and head. This is because the value for displacement of head and cylinder are unequal (note equations (7) and (11)). To overcome this difference, moment and shear forces are set up at the joint of head with cylinder wall which increase the stress values. So the total longitudinal stress is sum of the stress calculated for membrane only (5) and stress developed in the joint. In reality this stress is localized at the joint region only and away from the center only membrane stress needs to be calculated. Bending hoop stresses are much smaller and can be ignored.

4. Dimensional analysis of pressure vessel

4.1. Dimensional analysis
The material used for the pressure vessel considered is Maraging 250 Steel. The material and geometrical properties of the considered pressure vessel are given in Table 1. The results show that the stress scale factors are independent of the material properties and only strain and displacement scale factors are affected. Displacement and strain scale factors change by the same order of magnitude by which the model material deviates from the prototype. However, this entire study comes under the domain of partial similarity which is beyond the scope of this paper. Due to which, the results have not

IOP Conf. Series: Materials Science and Engineering **642** (2019) 012004 doi:10.1088/1757-899X/642/1/012004

been included in this paper. The thickness is calculated using ASME code UG-23. After calculation and allowances, thickness of 6 mm is selected. The general similitude theory for the pressure vessel can be found from the Buckingham pi theorem by considering factors that affect the vessels structural behavior. The validation may then be carried out by checking that the model and prototype pi products are equal.

14 physical parameters as shown in Table 2 are considered for the mentioned pressure vessel. These are 'σ_h', 'σ_l', 'σ_r', 'p', 'R', 't', 'υ', 'h', 'a', 'b', 'Δr', 'Δz', 'Q' and 'E'. All these variables can be represented by two dimensions which are force and length. As a result 12 dimensionless π products are left.

Table 1: Material Properties for Maraging 250 Steel and Pressure Vessel Dimensions

σ_y (yield strength)	1560 MPa
σ_u (ultimate tensile strength)	1760 MPa
E (Young's modulus of elasticity)	190 GPa
v (Poison ratio)	0.3
p (Design pressure)	5 MPa
R (external radius)	1000 mm
h (length)	6300 mm
Head ratio	2:1
Semi-major axis (external)	1000 mm
Semi-minor axis (external)	500 mm
Straight flange	40mm

Table 2: Physical Parameters and their Dimensions

Dimension	Force (F)	Length (L)
σ_h	1	-2
σ_l	1	-2
σ_r	1	-2
p	1	-2
R	0	1
b	0	1
Δr	0	1
t	0	1
v	0	0
E	1	-2
h	0	1
a	0	1
Δz	0	1
Q	1	-2

4.2. Estimation of dimensionless groups

The physical relation may be expressed as a relation of following 12 dimensionless π products, i.e:

G(π1,π2,π3,π4,π5,π6, π7,π8,π9,π10,π11,π12) = 0

π1 = (p,R,$\boldsymbol{\sigma_h}$) π2 = (p,R,$\boldsymbol{\sigma_l}$) π3 = (p,R,t) π4 = (p,R,υ) π5 = (p,R,E) π6 = (p,R,h)

π7 = (p,R,$\boldsymbol{\sigma_r}$) π8 = (p,R,a) π9 = (p,R,b) π10 = (p,R,Δr) π11 = (p,R,Δz) π12 = (p,R,Q)

The method of obtaining the first *pi* product is shown here. Rest of the *pi* products are obtained by using a similar method.

$$\pi 1 = p^a R^b \sigma h$$
$$\pi 1 = (FL-2)a(L)b\ (FL-2)$$

For

$$F \geq a+1=0 \tag{12}$$
$$L \geq -2a+b-2=0 \tag{13}$$

From equation (12) and (13) we get, $a = -1$ and $b = 0$

So the first *pi* product is: $\pi 1 = p^{-1}R^0\sigma h$. Rest of the *pi* products determined are:

$\pi 1 = \dfrac{\sigma h}{p}$ $\pi 2 = \dfrac{\sigma 1}{p}$ $\pi 3 = \dfrac{t}{R}$ $\pi 4 = \upsilon$ $\pi 5 = \dfrac{E}{p}$ $\pi 6 = {}^{h}/_{R}$

$\pi 7 = \dfrac{\sigma r}{p}$ $\pi 8 = \dfrac{a}{R}$ $\pi 9 = \dfrac{b}{R}$ $\pi 11 = \dfrac{\Delta r}{R}$ $\pi 10 = \dfrac{\Delta z}{R}$ $\pi 12 = \dfrac{Q}{R}$

12 pi-products obtained can be divided into two groups, dimensional group and pressure group as shown in Table 3. The dimensional group contains pi-products that are ratio of dimensions and displacements of pressure vessel. The pressure group contains products related to modulus and stresses.

4.3. Estimation of scaling factors

As there are two repeating variables so we can fix two scale factors and find scale factors for the rest of the vessel parameters using the calculated *pi* products. Scale factors chosen are λp and λR. Subscript 'pr' is for prototype and 'm' for scaled model.

$$\lambda p = \frac{(p)_{pr}}{(p)_m} = 5 \text{ (Pressure scale factor)}, \qquad \lambda R = \frac{(R)_{pr}}{(R)_m} = 10 \quad \text{(Radius scale factor)}$$

Similitude demands that *pi* products of prototype and model remain equal therefore;

$$\frac{\sigma_m}{p_m} = \frac{\sigma_{pr}}{p_{pr}}, \qquad \frac{\sigma_{pr}}{\sigma_m} = \frac{p_{pr}}{p_m} = \lambda_p = \lambda_\sigma$$

$\lambda_{\sigma h} = \lambda_p = 5$ (Hoop stress scale factor), $\lambda_{\sigma l} = \lambda_p = 5$ (Longitudinal stress scale factor)

$\lambda_{\sigma_r} = \lambda_p = 5$ (Radial stress scale factor), from 3rd pi term $\lambda t = \lambda R = 10$ (Thickness scale factor)

and 5th pi term gives $\lambda E = \lambda p = 5$ (Youngs modulus scale factor) , $\lambda R = \lambda h = \lambda a = \lambda b = \lambda Q = 10$

where the scale factor of λh is for length, λa is for ellipsoidal head's semi-major axis, λb is for ellipsoidal head's semi-minor axis and λQ is for shear stress. $\lambda v = 1$ that is poisson's ratio for prototype = poisson's ratio for model. The scale factors found are summarized in Table 4 where $\lambda p=5$ and $\lambda R = 10$.

Table 3: pi Products Groups

Dimensional group	Pressure group
Π3	Π1
Π4	Π2
Π5	Π7
Π6	Π12
Π8	
Π9	
Π10	
Π11	

Table 4: Scale Factors Found by Applying Similitude Theory

Scale factor	Value	Scale factor	Value
λ_{σ_h}	5	λ_t	10
λ_{σ_l}	5	λ_h	10
λ_{σ_r}	5	λ_a	10
λ_Q	5	λ_b	10
λ_E	5	λ_υ	1

Using these scale factors scaled model dimensions and scaled loads are found next given the dimensions and applied loads on the prototype. Similitude between the prototype and its scaled model exists when all the pi terms are same for both the model and prototype. The tabulation of prototype and 1/10th model of pressure vessel is shown in Table 5 and the pi terms obtained for both the model and prototype is shown in Table 6. It is evident from the Table 6 that pi-terms for both the prototype and model are identical. Thus pi-terms shown in mentioned table have been verified for complete similarity. The pi-products verified belong to the dimensional group and as both prototype and model dimensions are available, they can be verified. Only one of the pi-product in the pressure group has been verified. For the verification of remaining 6 pi-products ANSYS is used. The structural analysis is performed for both the model and prototype and then model results are used to predict prototype results.

Table 5: Prototype and Model Dimensions

Parameter	Prototype	Scale factor	Model
Radius, R	1000 mm	10	100 mm
Height, h	6380 mm	10	638 mm
Thickness, t	6 mm	10	0.6 mm
Semi major axis(external)	1000 mm	10	100 mm
Semi-minor axis(external)	500 mm	10	50 mm
Pressure, p	5000 kPa	5	1000 kPa
Young's modulus, E	190 GPa	5	38 GPa

Table 6: Pi terms for model and prototype

Pi-products	Expression	Numerical value
$\pi 3$	(t/R)pr	6/1000= 0.006
	(t/R)m	0.6/1000= 0.006
$\pi 4$	(v)pr	0.3
	(v)m	0.3
$\pi 5$	(E/p)pr	190e09/5e06 = 38000
	(E/p)m	38e09/1e06 = 38000
$\pi 6$	(h/R)pr	6300/1000 = 6.3
	(h/R)m	630/100 = 6.3
$\pi 8$	(a/R)pr	1000/1000=1
	(a/R)m	100/100=1
$\pi 9$	(b/R)pr	500/1000=0.5
	(b/R)m	50/100=0.5

5. FEA (Finite Element Analysis)

The geometry is made in PRO/E as shown in Figure 1 and then imported into ANSYS software. An axisymmetric FE model of the scaled pressure vessel is developed in the same way as the prototype. The requirement is to map the results found in ANSYS of the scaled model of the vessel onto the prototype by using the derived similarity parameters called scale factors and compare them to the ANSYS results of the actual prototype. After some preliminary treatment of the ANSYS result data, both the prototype and model data is compiled in the same M-file of MATLAB program. In each M-file the scale factors derived earlier using similitude theory (scale factor of 5 for stresses and 10 for geometrical parameters respectively) are used to map model stress and strain values on to the prototype. This process basically predicts prototype results using model results and similarity parameters. These mapped results are referred to as predicted prototype result from now on to avoid confusion.

5.1. Analysis of results

The structural parameters analyzed for prototype and model are Von Mises stress, Radial stress (σ_r), Longitudinal stress (σ_l), Hoop stress (σ_h), Shear Stress, Q in rz plane (XY when taken in terms of cartesian coordinates), Radial displacement Δr (X coordinate displacement), Longitudinal displacement Δz (Y coordinate displacement), Radial strain and Longitudinal strain. The observation made from the comparison of prototype and predicted prototype results are summarized below.

The Von Mises stress has been used as failure criteria. Figure 2 shows the maximum Von Mises stress to be about 1300MPa which is less than the yield stress of Maraging 250 Steel (1560MPa). The vessel is not in a state of failure. The same method can be used for any other material using its respected properties. However, the limitation is to choose a material for the scaled model as per the scaled down properties.

IOP Conf. Series: Materials Science and Engineering **642** (2019) 012004 doi:10.1088/1757-899X/642/1/012004

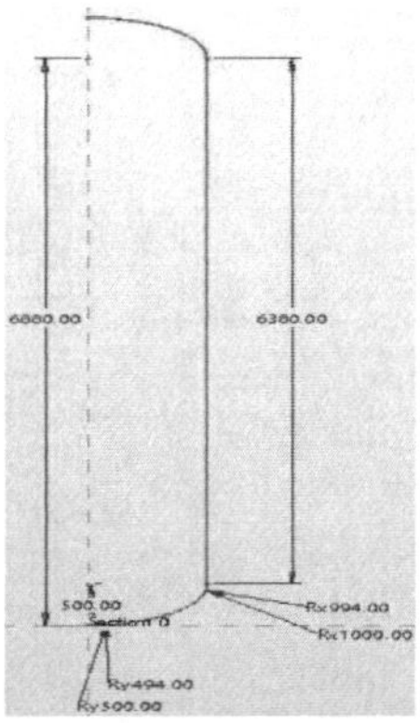

Fig 1. Pressure vessel geometry

Table 7: Percentage errors in analyses results

Parameter	Percentage error
Von Mises stress	<1.4%
Radial stress	< 0.1 %
Longitudinal stress	< 1 %
Hoop stress	< 1 %
Shear stress	< 0.35 %
Radial displacement	< 0.5 %
Longitudinal displacement	< 0.1 %

The predicted and actual prototype results match for all the stresses and displacements as is evident from the overall percentage errors shown in Table 7. Percentage error is insignificant and maximum values of error occur in the heads. However the error magnitude is negligible when compared to the actual values. For radial, longitudinal and hoop stress, the maximum percentage error is less than 1% for the stresses due to the change in the geometry from cylindrical to the head region. The percentage error in radial and longitudinal displacement is less than 0.5%.

During development of similitude theory, shear is taken as a parameter as it is created at the joint of pressure vessel. The graph in Figure 3 of shear stress results shows that it is almost zero throughout the cylindrical region but has notable presence in the heads which proves that it should be included as a parameter during dimensional analysis. The strains are not chosen as a parameter because it is dimensionless and its pi-products will always be 1(i.e. model and prototype have same strain). The model strains have been plotted as it is (not multiplied by scale factor) along with the prototype. As shown in Figure 4 and Figure 5, the two map onto each other verifying that strains are not a similitude parameter.

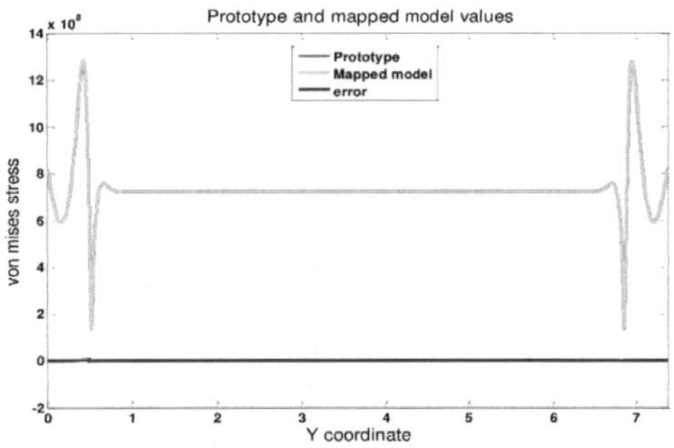

Fig. 2: Prototype, predicted prototype and error plots for von mises stress

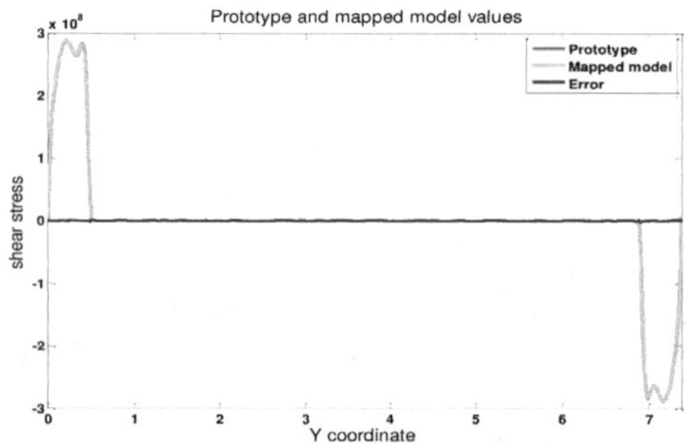

Fig. 3: Prototype, predicted prototype and error plots for shear stress

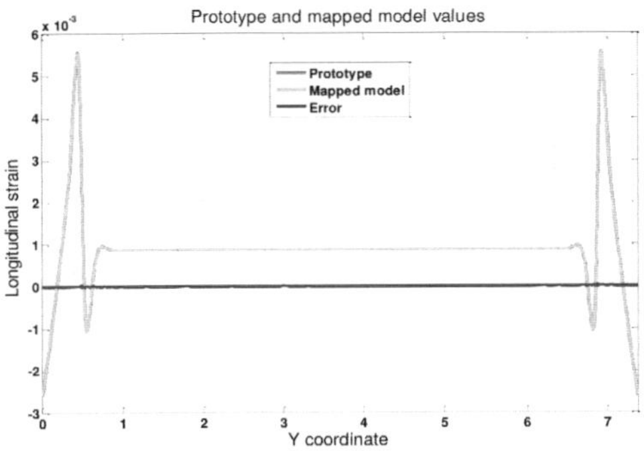

Fig. 4: Prototype, predicted prototype and error plots for longitudinal strain

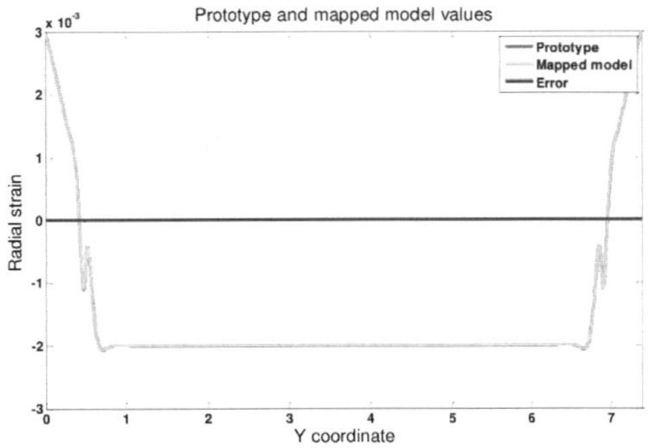

Fig. 5: Prototype, predicted prototype and error plots for radial strain

6. Conclusion

Using similitude theory, a 1/10th scaled down pressure vessel, which duplicates full scale vessel, has been analyzed and all pi products for the two are verified. Analysis of the results shows that the percentage error between the prototype and mapped model is less than 1.5% for stresses and 0.5% for displacements respectively. This guarantees that scale factors determined using the established similitude theory, can be used to make a scaled model which accurately predicts prototype response of the pressure vessel. Theory of similitude is a very useful tool as it predicts response without developing full scale models. It is very economical where large scale testing is required. It gives the rules that should be followed, if complete prototype results are to be predicted using a scaled model

Acknowledgement

The authors would like to acknowledge Dr. Suleman Pasha for his guidance and motivation throughout the research.

References

[1] Ramu M., Raja V. P., Thyla P. R.. Establishment of Structural Similitude for Elastic Models and Validation of Scaling Laws. KSCE Journal of Civil Engineering 2013; Volume (17): 139 – 144.

[2] Coutinho C. P., Baptista A. J., Rodrigues J. D.. Reduced Scale Models Based on Similitude Theory: A Review up to 2015; Engineering Structures 2016; Volume (119): 81-94.

[3] Zhu Y., Wang Y., Luo Z., Han Q., Wang D.. Similitude Design for the Vibration Problems of Plates and Shells: A Review. Frontiers of Mechanical Engineering 2017; Volume (12): 253-264.

[4] Fang B., Jin Z., Li H.. Theoretical Research and Experimental Verification on Vibration and Acoustic Similarity of Stiffened Cylindrical Shells. In 2018 7th International Conference on Energy and Environmental Protection (ICEEP); Atlanis Press.

[5] Wang J., Li L., Z., Yu W.. Structural Similitude for The Geometric Nonliniear Buckling of Stiffened Orthotropic Shallow Spherical Shells by Energy Approach. Thin Walled Structures 2019; Volume (138): 430-457.

[6] Simitses G.J., Rezaeepazhand J. Structural Similitude and Design of Scaled Down Laminated Models. NASA Technical Report December 1993.

[7] Zhou L. L. and Li D. K.. Design of Scaled Model for Dynaic Characteristics of Stiffened Cylindrical Shells Based on Equivelent Similar Method. The Aeronautical Journal 2019; Volume (123): 398-415.

[8] Nicoreac M. P., Pârv B. R., Petrina M., Petrina T. Similitude Theory and Applications. Acta Technica Napocensis: Civil Engineering & Architecture 2010; Volume (53): 187 – 196.

[9] Anderson J. D. Fundamentals of Aerodynamics. 3rd Edition. Mc. Graw Hill 2001: 31-32.

[10] Megesy E. F. Pressure Vessel Handbook: 13-23.

[11] Meyer J. E. Structural Design Notes, Topic C Pressure Vessel Stress Analysis. Rev: Aug 1996.

[12] Warren C Young. Roark's Formulae for Stress & Strain", 6th Edition. Mc. Graw Hill 1989:518-519.

Stabilizing Attitude Control For Mobility Of Wing In Ground (WIG) Craft - A Review

Matdaud, Z.[1]*, Zhahir, A.[1]*,Afifi.A Pua'at.[1]*, Aziz Hassan[2]* , Ahmad, M.T.[3]*

[1]Department of Aerospace Engineering, Faculty of Engineering, Universiti Putra Malaysia (UPM), 43400 Serdang, Selangor Darul Ehsan, Malaysia.
[2]Retired Professor in Mechanical Engineering (Automotive), Universiti Teknologi Malaysia.
[3]Centre for Defence Research and Technology (CODRAT), Universiti Pertahanan Nasional Malaysia (UPNM), Kem Sungai Besi, 57000 Kuala Lumpur, Malaysia.

*Corresponding author e-mail: zmatdaud@gmail.com, amzari@upm.edu.my

Abstract. Wing in ground crafts development have been rapidly advancing in recent years. The current paper reviews the researches and developments of existing stability control system technology's development and enhancement for wing in ground effect crafts. The review is critically intended for the development of the control system for two-seater Dragonfly 2, a hoverwing type craft. The current review will commence with the introduction on the theory behind the in ground effect phenomenon on the crafts, its regulations and the types of wing in ground crafts, their advantages and disadvantages, and their stability and control issues. This paper also discusses the available attitude control sys-tem types of wing-in-ground craft (WIG), its experiments, simulations and computational methods done especially on both the lon-gitudinal and lateral motion stability.

Keywords: 2-Seater Hoverwing, Attitude Control, Flight Control, Wing in Ground Craft

1. Introduction

The study of the ground effect phenomenon and the construction of wing –in-ground (WIG) crafts have been going on for over half of a century since the Soviets/Russians proved the concept so dramatically through the Ekranoplan programs of the 1960 – 80's. A lot of researches are progressing while the present review is being written, which mostly deal with ground effect aerodynamics, maneuverability and hydrodynamics of the WIG crafts. Not until recently, only a few WIG crafts have been certified by International Maritime Organization (IMO) that the said crafts are a ground effect worthy.

Mentioned herewith are two of them, Airfish 8 and WSH-500. In 2010, Lloyd's Register has certified the 8-seater Airfish 8 (Figure 1-a.) refurbished by Wigetworks Pte. Ltd. of Singapore as the world's first WIG effect craft to fully surveyed to IMO Rules which the craft is manufactured in Germany and refurbished in Singapore [1]. Wing Ship Technology (WST) of Korea later claimed that their 50-seater WSH-500 (Figure 1-b.) is an A-type WIG which has been fully surveyed during the design and manufactured by Lloyd's Register and is certified to operate only in ground effect [2-3].

Fig. 1:-a. Airfish 8[4] **Fig. 1:-b.** WSH-500 [5]

These highlighted certifications proved that the interests in WIG crafts development are increasing at a very rapid pace and dramatic. These certifications prove that the long outstanding issues and limitations highlighted by The Aeronautical and Maritime Re-search Laboratory such as stability and control, aerodynamics analysis and systems, operation of aircraft structures in marine environments, and a relatively high maintenance cost engines and load bearing structures being exposed to corrosion by seawater mist [6] can be resolved with the emerging new researches, developments and technological advancement.

The new millennium witnessed many researches are being studied and papers are being published in the aerodynamic [7] and hydrodynamic [8] stability of WIG crafts, especially during the take-off, heave, pitch and surge motion compared to the transverse and lateral stability that includes the axis roll, yaw and sway attitudes [9]. In addition to dominant wind within the ground effect area of sea level, the ever unstableeffect of the wave current [10] is another concern for the instability caused by the aerodynamics of the wing of a smaller sized WIG crafts.

Moving from that point, Department of Aerospace Engineering of Universiti Putra Malaysia (UPM), together with the Centre of Defense Research and Technology (CODRAT) of Universiti Pertahanan Nasional Malaysia (UPNM) are collaborating on an investigation of developing the first ever in Malaysia a 2-seater hover-wing, a Type-A WIG effect craft namely Dragonfly DF-2, a WIG craft that is able to reduce the aerodynamic instability issues during take-off, in-flight, and landing motion. As a part of this investigation, the stability and maneuverability of the WIG craft are going to be the major focus and consideration of this review.

2. The Ground Effect Phenomenon

IMO defines ground effect as "*a phenomenon of increase of a lift force and reduction of inductive resistance of a wing approaching a surface*" [11]. The boundary of this ground proximity phenomenon is not fixed and it depends on the size, depth and span designs of the WIG craft. Typically, it occurs when a craft is at "*an altitude less than the mean chord length of the wing*" from sea surface or half of the wing span. Rozhdestvensky [12] defines that ground effect as "*an increase of the lift-to-drag ratio of a lifting system at small relative distances from an underlying surface*". Halloran & O'Meara [6] addressed that "*in terms of the total pressure of the flow, the additional lift is due to a rise in static pressure under the wing and causes a change to many of the other aerodynamic characteristics of the wing*" and this is illustrated in Figure 2. Yun et. al. [13] defines that ground effect is "*the enhanced lift force acting on a wing that is travelling close to the ground or water surface, commonly less than one wing chord height*" due to higher slow-down of the air trapped between the ground and under wing surfaces which generates greater increase in pressure on the under wing surface.

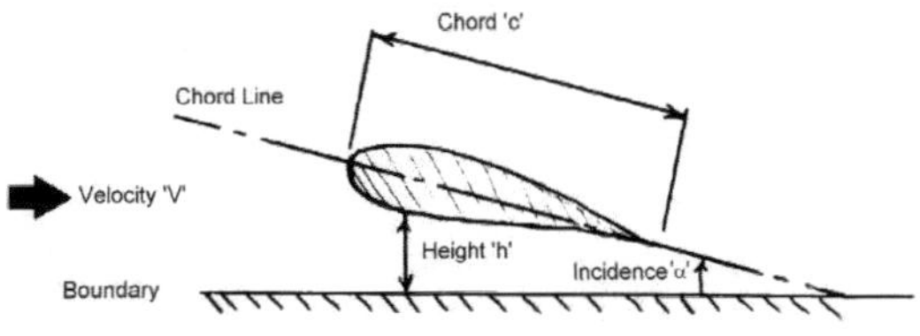

Fig. 2: Ground Effect on a Wing [6]

Aircraft pilots generally experience the similar ground effect during the landing mode, where extra lift is gained onto the wing prior to the lower part of the carriage reaches the ground. This phenomenon leads the aircrafts to glide above the runway until the cushion decays after sufficient loss of their speed. This is because the airflow between the aircrafts' lower part and the ground has been blocked such that the pressure on the under surface of the wing and the lift increases [13]. Furthermore, for a wing operating in ground effect zone, the downwash velocity will be reduced by wing tip vortices, and the induced drag reduced by the wing tip vortex-induced velocity [13] as shown in Figure 3. The ground effect reduces the down-wash velocity which causes an increase in the lift and reduces the drag, in conjunction with increasing the effective aspect ratio of the wing. These are the two most important features of WIG crafts.

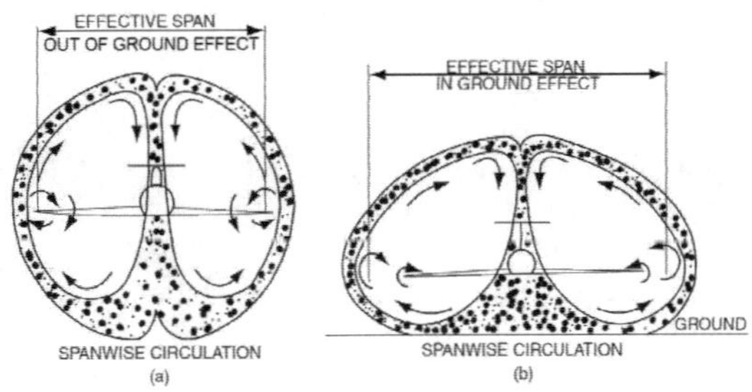

Fig. 3: Wingtip vortex difference without (a) and with (b) ground effect [32]

IOP Conf. Series: Materials Science and Engineering **642** (2019) 012005 doi:10.1088/1757-899X/642/1/012005

3. WIG Craft Definition, Classifications Regulations and Advantages

Wing in Ground Effect (WIG) term has been officially adopted by the "United Nations' International Civil Aviation Organization" (ICAO) and the "International Maritime Organization" (IMO) for a marine craft using ground effect as a means of lift [14]. According to IMO, "*WIG craft have much in common with aircraft characteristics that in their ground effect zone they are supported only by aerodynamic forces of the wing which enable them to operate at low altitude above the sea water surface without any contact*" [11].

Both IMO and ICAO have agreed that for any WIG craft that is capable of sustaining its flight outside of the ground effect zone shall be subjected to the rules and regulations of ICAO [11]. On the other hand, for other crafts which includes the limited "fly-over" capability crafts shall be subjected to the rules and regulations of IMO.

The IMO has categorised WIG craft types as follows in Table 1.

Table 1: The Classification of WIG Craft Operation by IMO [11]

WIG Category	Crafts Certified Operation	Regulatory Authority
Type A	Within crafts' ground effect zone of the sea surface	IMO
Type B	Momentarily increase crafts' height beyond ground effect zone but below 150 m above the sea surface	Both IMO and ICAO
Type C	Beyond crafts' ground effect zone and surpassing height of 150m above the sea surface	ICAO

Figure 4 below depicts an interpretation of the classification of a WIG craft.

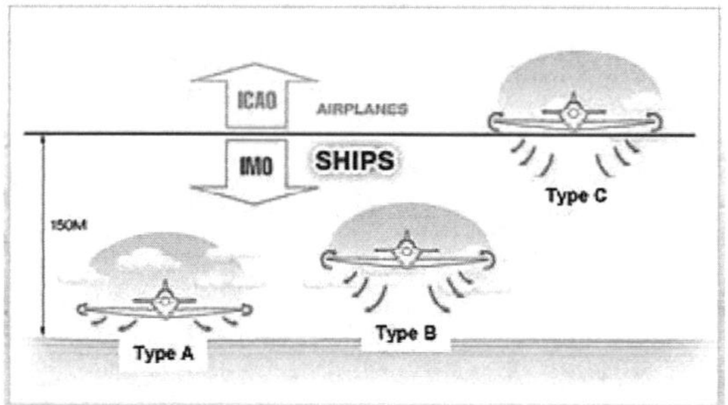

Fig. 4: The Regulatory Authority of Craft Types [15]

Rozhdestvensky [12] outlined that aerodynamically, WIG craft can be classified with the different configurations depending on the method used to satisfy the requirements of longitudinal stability. Table 2 shows the details of each configuration.

The Safety Committee of IMO in its 76th session has approved the Interim Guidelines of a WIG Craft in 2002 under MSC/Cir. 1054 [11]. The configuration of WIG craft falls between the maritime and aviation regulatory regimes in view of which the Interim Guidelines of a WIG craft were developed. These guidelines were intended for "*proper engineering analysis, design and developmental testing to achieve an adherently safe WIG craft to operate within the ground effect*". It is essential to note that WIG craft have much alike with aircraft characteristics, but it operates within the other waterborne vessels' zone as such the Interim Guidelines of a WIG Craft stated that its operation must comply and use the equivalent "collision avoidance rules" as of the typical shipping rules.

The guidelines were later amended in 2005 for the Wing-in-ground Craft Safety Certificate and Record of Equipment document [16]. In addition to the guidelines, the "*Sub-Committee on Standards of Training and Watch-keeping*" of IMO has also approved the "*General Principles and Recommendations for Knowledge, Skills and Training for Officers On Wing-In-Ground (WIG) Craft Operating in Both Displacement and Ground Effect Modes"* in MSC/Cir.1162, focusing to "primarily assist Member governments in developing their national requirements for qualification and certification of officers on a WIG craft operating in both displacement and ground-effect modes" [17].

IOP Conf. Series: Materials Science and Engineering **642** (2019) 012005 doi:10.1088/1757-899X/642/1/012005

Table 2: The Configuration of WIG Craft [12]

WIG Configuration	Main Design Feature	Advantages	Disadvantages	Models Built
Tandem	- Adjusted pitch angles design - Adjusted wing elements design (fore and aft) geometry	- Construction: simple - Configuration for static stability margin tuning: simple - Control: throttle (one channel), effective - Span: small	- Operation mode: ground effect zone (GEZ) only - Marginal static stability: very sensitive to pitch angle and ground gap combinations	SM-1, Tandem Aerofoil Boat (TAB)
Airplane-type wing-tail	- Lower main wing (closer to ground) - Mounted tail plane (horizontal type) on vertical stabilizer - Height is stabilizer outside ground effect zone	- Stable craft flight sustainability: good - Heights and height-pitch combinations range: large - Power enhancement at take-off: possible to efficient - High speed (wing loadings are large) - Turning manoeuvres: efficient (use "hop" and banking ability)	- Weight fraction (empty): large - Weight penalty: very large - Structural weight: large Note: (structurally large craft and have large tail unit highly mounted)	Kaspian Monster, Orlyonok, Loon, Strizh,
Airplane-type wing-tail (Reverse Delta Planform)	- Used reverse delta planform type for main wing - Tail unit is quite small relative to wing	- Lift-Drag (L/D) ratio: high - Stable flight heights and pitch angle range: large - Capable to do "dynamic jump" manoeuvre - Turning manoeuvres: efficient	- Power augmentation: absent - Take-off aids: Inefficient - Overpowering	Lippisch-type craft (X-112, X-113, X-114) Airfish craft family, Eska, XTW
Flying wing	- Reduced components (not related to increase lift) - Smaller tail (horizontal type) or no tail	- Maximum advantage of ground effect (GE)utilisation: efficient - Weight fraction (empty): low	- Height-pitch combinations range: low - Flaps control: inefficient	Amphistar Aquaglide WISE KAG-3
Composite wing	- Centro plan wing type with endplates (aspect ratio: small), and side wings (aspect ratio: high) - Under-wing (profiled to make tail unit smaller)	- L/D ratio: Higher - Efficiency and range: Higher - Efficiency of the power-augmented take-off: Maximized		Ekranoplan (MPE)

Many studies have addressed that there are advantages and yet dis-advantages to the operation of a WIG craft. Foremost were its range and payload capabilities. Since the WIG crafts are designed to fly a few meters above water by making use of the ground-effect, they have the technical advantages above traditional airplanes and vessels because they are more energy efficient in their speed range. Com-pared to conventional marine craft vessels, WIG is significantly faster [6].

Considering the fact that WIG crafts exploit the ground effect zone during their cruise mode rather than cruising on the seawater, it can be easily understood that they have higher lift to drag ratio giving them the potential for greater efficiencies than even aircraft [18,12]. The WIG crafts will achieve an increase in its maximum speed and optimal cruise speed with reduction in drag or increase in lift to drag ratio [6]. According to Yun et. al. [19] the increased lift to drag ratio of WIG crafts due to the sea surface effect may result in they are being more economical compared with aircraft. The ekranoplanes have the ability to carry the payload of a very large weight and overall dimensions at a farther range [18], have better operational cost efficiency than aircraft [6,8,12-13] and offer their passengers better quality services [8] as such the comfort level is as close as to vessels standard once they cruise in the air [18].

Operationally, the WIG crafts have good potential to serve as military crafts for their ability to operate below radar detect ability and near coastal areas [6,12]. Nebylov [18] outlined that WIG crafts have the upper hand on their nonessential of special aerodromes since passengers boarding can be implemented directly in a central part of a seaport and they are also exploitation-safe crafts due to their capability of cruising at a very low altitude and landing on water. Park et. al. [8] stated that WIG crafts consume fewer fuel than other high speed vessels including hydro-foil craft effect and they do not require expensive infrastructure because of their ground effect principle exploit. Rozhdestvensky [12] outlined the classification of WIG crafts highlighting its advantages and disadvantages as in Table 2.

Apart from the Interim Regulation by IMO in 2002, there is no consensus for longitudinal stability control [7]. By the time the study was made by [20], they indicated that WIG crafts have never reached a satisfactory reception as mainstream marine transport crafts. According to [21], the commercial competitiveness of WIG craft has been weakened by tremendous impact of the maximum price of WIG craft onto its direct operating costs. Aforementioned licenced crafts may not wholly have accepted by the investors for that particular reason, but others are still intended for hobbyist, demonstrators or test flight crafts. Wiriadidjaja et. al. [7] claimed that system life cycle model of WIG craft is stagnant at the infant development, still has not evolved to its crawling stage, but rather a mere conceptual development stage.

4. The Importance of WIG Stability and Attitude Control and Issues

Many studies have been focusing on the WIG craft stability in specific areas during its take-off, cruising in ground effect and landing onto the sea surface recently. The interim guide from the IMO re-quires that; any due certified WIG craft should have at least controller of suitable design for lateral, longitudinal, and attitude control of the WIG craft [11]. These WIG craft stability controls can be achieved by using "*air or water rudders, foils, flaps, propellers or jets which may be steerable, yaw control ports or side thrusters, differential propulsive thrust, variable geometry of the craft or its lift-system components or by a combination of these devices*" [11].

In addition to the recommendations, IMO has also outlined that the on-board officers of a WIG craft must be qualified and certified with adequate training in maritime and aviation knowledge and skills of handling the unique complexity of a WIG craft [17]. These requirements are somewhat general for WIG manufacturers and operators as a pre-requisite safety front for commercial operation to make sure that there is no fatalities and subsequently impair or cap-size the craft [8].

The flights of WIG crafts and their stability control are very challenging as such these issues are the most likely why WIG craft stability-related safety issues is still being studied by many scientists and engineers to solve for in the wake of commercial operation of WIG crafts. Unlike seaplanes, WIG crafts take-off from the sea surface and cruise at a constant gap just a few meters from the sea surface, and of course the significant landing on the sea surface.

The state of sea surfaces is naturally propagating waves in the presence of wind. However, dominant and strong wind may cause significant and rough wave condition on the sea surface [9]. This rough wave condition causes disturbances to the operation of WIG craft, may it be during take-off, cruising flight or landing mode. With the presence of these disturbances, stability problems arise that make safety issues of operating WIG crafts become more delicate to handle. Thus, the operation of the WIG crafts that are close to the sea surface at high speed requires a good stability in both longitudinal and lateral directions [22].

Staufenbiel & Schlichtingt [23] analysed the stability of an aircraft in longitudinal motion during its flight in ground effect and observed that the ground proximity has caused variations in the aircraft's longitudinal stability that were responsible for significant changes in the landing course, particularly whilst the flare-out manoeuvres. As aforementioned in Table 1 [12] Rozhdestvensky has aerodynamically classified WIG crafts as the different configurations depending on the method used to satisfy the requirements of longitudinal stability. These classification of WIG crafts have their own advantages and disadvantages that affected the longitudinal stability of the crafts.

Disturbances affect the longitudinal stability such that establishing a control system that can stabilize the WIG craft is essential [10]. WIG crafts are concluded to be safely operated when cruising flight is at maximum reliability, as such the trailing edge of the wing do not touch the water surface [24], or the disturbances. The disturbance issues are the non-linear aspects of the aerodynamics of ground effect zone, thus Hahn et. al. [24] considered that advanced control system concepts are highly demanded for to overcome the WIG crafts hindrance for actuators and states.

The hydrodynamic drag during take-off the most significant drag that contributes the problem on resolving the power demand and distance for take-off and landing [25]. Power Augmented Ram (PAR) is one of the devised and tested lifting power scheme that enhances the air cushion under the wing by blowing high pressure air which can give the WIG craft the high lift at low speed they require during take-off and landing. Another essential approach to improve WIG craft performance is using hydrofoil which has similar shape and principle to aerofoil used by aircrafts, particularly to reduce the hydrodynamic drag during the planning mode of the WIG craft and substantially gains speed and lift during take-off [25].

WIG crafts also experience natural aerodynamics roll angle of stability in lateral aspect which is considered as an remarkably significant aspect for safety of their flight [26-27]. In addition, Amir et. al. [9] highlighted that to avoid unforeseen accidentals, a thorough understanding on the transverse or lateral motion stability is crucial and; roll, yaw and sway are typical dynamics to WIG crafts during their flight cruising phase because of the wind on the sea surface is more dominant compared to an altitude higher than the ground effect zone.

5. Significant WIG Crafts Control System De-velopment Regarding Stability and Attitude Control Issues

Detailed submission guidelines can be found on the journal web pages. There are studies done that investigated regarding WIG crafts stability which are going to be reviewed later in this chapter, and Amir et. al. [22], Kornev & Matveev [26] and Divitiis [28], to name a few, are studies that have been done in term of attitude control of a WIG craft. Successful WIG craft models have been running their trials all over the world for years, yet commercialisation is still at halt. This situation is portraying a somewhat grey impression towards the possibility of the craft could finally accepted by the marine industry players. Safety, among the reasons, is the key factor that needs special attention Park et. al. [8] to convince the marine industry players to invest in its customer attraction.

From a safety viewpoint, it is necessary to avoid accidentals and crashes. While performing a banked turn, the risk of the wingtip of a WIG craft touching the ground surface is very high. Kornev & Matveev [26] highlighted that depending on the roll angle, the WIG craft gains additional aerodynamic forces and moments while per-forming a banked turn at the ground proximity. This manoeuvre must be performed with a limited roll angle at a safe height between the wingtip and the sea surface to avoid the touchdown risk. This limits the manoeuvrability of the WIG craft as it demands a very large turning

IOP Conf. Series: Materials Science and Engineering **642** (2019) 012005 doi:10.1088/1757-899X/642/1/012005

radius [26]. This phenomenon has been investigated by Divitiis [28] that the computational analysis for the banked turn explained the behaviour observed in ground effect, while the stability analysis provides clarification about the flying characteristics with appropriate respect to the spiral mode stability and oscillating rolling mode nature.

Wave disturbances are another safety concern which are very hard to predict thus lead to number studies were conducted either experimental or by simulation. Nebylov et. al. [10] in their research have found the effective way of minimising the wave disturbance problem by optimising the landing approach direction of WIG crafts in relation to a general direction of sea waves' propagation, and have developed technology application which made the operation of WIG crafts on all-weather conditions allowable. Therefore, Nebylov et.al. [10] suggested that during landing, the distinct function of minimising mechanical loadings would accomplish the coordination control of all steering-related structure.

In another paper a year later, Nebylov & Nebylov, [29] discovered "*the laws for change of all rudders positions*" *as such that* "*the pitch angle and engines thrust should be adhered to the most prob-able time of landing of the first contact between WIG craft and the disturbed sea surface*". They elaborated that all steering related structures of the WIG craft shall be coordinated by an on board control system as such the information of sea waves properties can be obtained prior to landing as shown in Figure 5. They suggested that all flaps and elevator shall be implemented foremost [29].

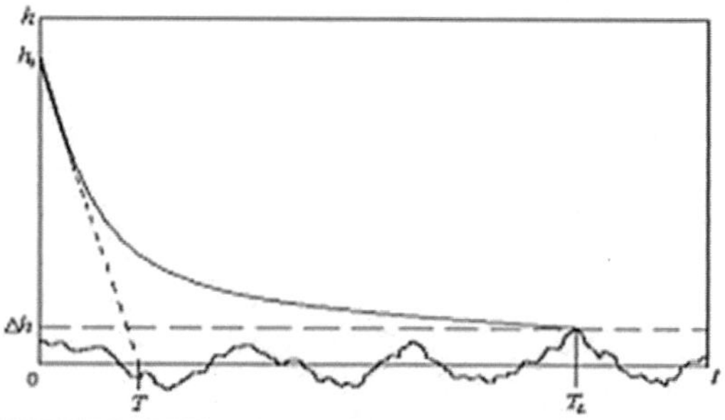

Fig. 5: The proposed "Law of Height Change in The Final Stage of Landing" by [29]

As it been known that the ground effect action is the most significant when the WIG flight is at an altitude of less than the mean of the wing chord as presented in Figure 6. Thus, the possibility of the action at the limited height of sea waves is dependent a particular size of WIG craft. In order to fulfil the safety of flight at a certain permissible height of sea wave's criterion, it is imperative to select the flight at an extremely low altitude. Daniel [30] suggested that automatic height control system establishment can assure the functional characteristics required for rough sea circumstances even though if the craft has the ability to perform altitude and angle inclination self-positioning naturally, and its necessity to provide good flight stability in the case of overcoming the essential longitudinal plane motion instability of particularly WIG crafts.

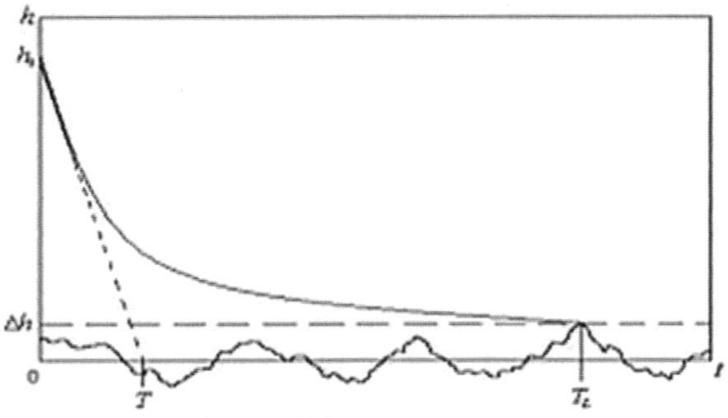

Fig. 6: Coefficient of lift vs. height/chord (h/b) [30]

To achieve a safe low altitude flight, a particular Lift-to-Drag ratio and stable flight are required. Daniel [30] implemented a control system that has the capability to automatically maintain the WIG craft at an ideal flying altitude as his objective. He adopted the concept of a system that can take control the elevator along with ability to rectify the flying altitude while in the initial take off until the craft reached its mean chord height and maintain it there whilst providing accurate flight altitude data. As compared to aircraft monitoring system, this significant insight sets a new standard for operating WIG craft that requires various hardware components such as actuators and sensors, algorithm control logic programming and the flight motion dynamics understanding as illustrated in Figure 7.

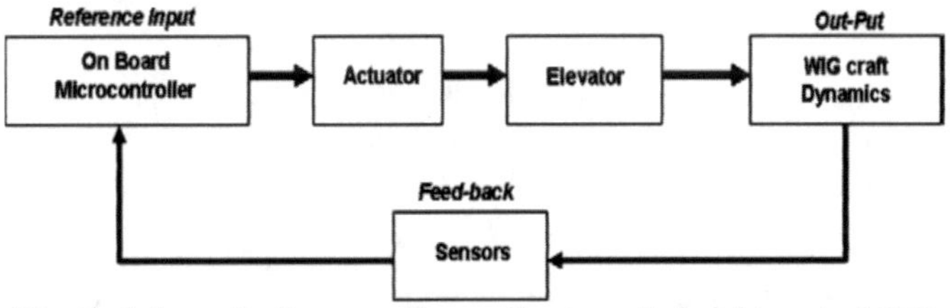

Fig. 7: Schematic diagram of overall automatic height control [30]

It is worth noted that the aforementioned findings have been updated and there are two concepts established by Nebylov et. al. [31]. Firstly, during its flight, a WIG Craft can be stabilised with respect to the average level of disturbed sea (Figure 8-a) and the craft flies in a straight path above the crests of waves with a clearance margin of Δ. Their insights have the ability to avoid the collision with the wave's crests even if there are some errors in controlling the motion. They suggested that there is a possibility with a very ideal altitude stabilisation system, reducing the clearance margin can in-crease the craft's flight efficiency [31].

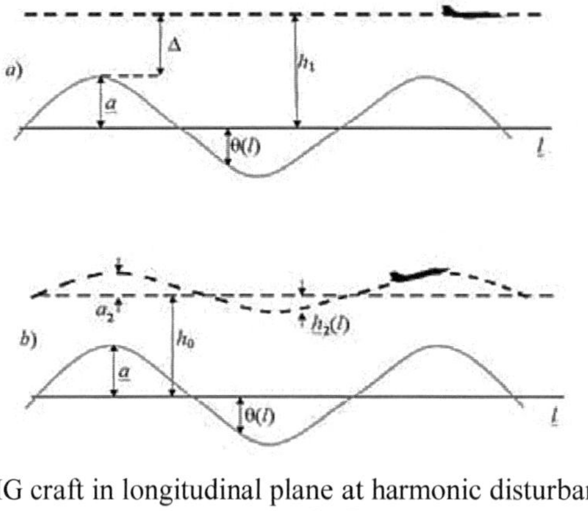

Fig. 8: Flight course of WIG craft in longitudinal plane at harmonic disturbance: a) 1st concept, b) 2nd concept [31].

The stabilisation of altitude of flight in longitudinal plane is slightly easier for the second concept (Figure 8-b). It permits the craft to capture partly the signal from the harmonic disturbance passing through the motion sensor equipped with transfer function capability. The flight is said to be of more effective as the performance of the synchronised phase is much reasonable to acquire from the motion control system of the IWG craft given that it uses the radar powered wave disturbance predictor or estimator [31].

Wang et. al. [32] numerically analysed the design criteria (wing anhedral, winglet, and tail) positioning on a conceptual WIG Craft design. They found out that somehow the aerodynamics efficiency and stability of the craft are effected by the positioning combinations of the design criteria [32]. Therefore, the selection of suitable combination of design criteria would result in compromising the aerodynamic efficiency and stability during the ground effect.

Hydrodynamics and aerodynamics were the major considerations for Yang et. al. [25] whose on their paper proposed the mathematical model for WIG craft in the longitudinal motion. They analysed the equations and criterion for static stability and discovered that the hydrodynamic centre in lift should be located both at the down-stream of the aerodynamic height centre, and the downstream of the WIG craft's centre of gravity in water. They also suggested that the hydrofoil should be located in front of WIG craft's centre of gravity to provide the static stability in longitudinal motion during take-off and cruising in ground effect. They also highlighted the aforementioned aerodynamic height centre should be at the upstream of the WIG craft's aerodynamic pitch centre while cruising in ground proximity region [25]. These insights are very valuable for WIG craft developers as such the knowledge can be useful to analyse as the design phase is concerned.

On the lateral stability of WIG crafts, previous study found that the lateral mode of oscillation for a single wing was stable in both roll and yaw, the latter being subject to fin and endplate configurations [33]. In a computational analysis done by Divitiis [28], the attitude of the flight significantly modifies the lateral stability of the craft. In this lateral state, WIG crafts have natural aerodynamic stability of roll angle in flight ground effect. The effect of heeling angle being subject to NACA 6409 wingtip at 13 degrees anhedral angle result-ed that when the WIG craft side heeled over to a precarious angle, the ground effect increases the side's vertical lift thus pressure distributions provide a natural righting moment when the WIG craft heels near sea water which causes a generation of recovering trans-verse moment [22].

In compliance with the IMO Interim guidelines, Wiriadidjaja et. al. [7] indicated that the stability augmentation capability features together with an efficient design integrated with an automatic control system of the WIG craft shall result in an inherently safe and reliable aerodynamic features. These features will also establish accurate flight data for WIG craft navigation management while reducing the pilot's workload.

6. Conclusion

Wing-in-ground and hoverwing crafts development have been rap-idly advancing in recent years in term of stability, mostly longitudinal. The current paper has reviewed the theory for the in-ground effect (IGE) phenomenon on the crafts, regulations and types of WIG crafts, their advantages and disadvantages, previous and re-cent researches and developments of existing altitude and attitude control technology enhancement for wing-in-ground effect and hoverwing crafts. There are many configurations that can be considered for the WIG effect craft Dragonfly DF2, as such the WIG craft that is able to reduce the aerodynamic instability issues during take-off, in-flight, and landing motion. The insights from the re-viewed paper will be based on the inventing the latest control system suitable for a small WIG especially on the longitudinal and lateral motion stability.

Acknowledgement

The authors would like to thank Universiti Putra Malaysia for providing financial support for the preparation of the article in the writing of the report and in the decision to submit the paper for publication.

References

[1] Wigetworks.com. Airfish 8 IMO 100A1 Certification by Lloyd's Register. 2010; Retrieved October 21, 2018, from http://www.wigetworks.com/w-w/wp-content/uploads/LloydRegisterGroup-Aug2010.pdf.

[2] Lee, H. J., Kang, B. J., Park, J. H., Lee, C. M., Kang, K. J., & Kang, C. G. Development of Hoverwing Type WIG Craft WSH-500. IEEE 2011: 1–5.

[3] Wingship.com. WSH-500 Design and Manufacturing Certification by Lloyd's Register. 2013; Retrieved October 21, 2018, from http://www.wingship.com/?page_id=275#.

[4] Wigetworks.com. Airfish 8. 2010; Retrieved October 21, 2018, from http://www.wigetworks.com/airfish-8/

[5] Wingship.com. WSH-500. 2013; Retrieved October 21, 2018, from http://www.wingship.com/?ckattempt=3

[6] Halloran, M., & O'Meara, S. Wing in Ground Effect Craft Review. DSTO Aeronautical and Maritime Research Laboratory 1999; doi: DSTO-GD-0201.

[7] Wiriadidjaja, S., Zhahir, A., Mohamad, Z. H., Razali, S., Pua'at, A. A., & Ahmad, M. T. Wing in Ground Effect Craft: A Case Study in Aerodynamics. International Journal of Engineering & Technology 2018; 7: 5–9. doi: 10.14419/ijet.v7i4.13.21319.

[8] Park, J. H., Kim, D., & Kang, M. Safety Requirements for Wing-In-Ground (WIG) Craft. IEEE 2011; 8 (1). doi:10.1109/OCEANS-Yeosu.2012.6263640.

[9] Amir, M. A. U., Maimun, A., Mat, S., Saad, M. R., & Zarim, M. A. Wing in Ground Effect Craft: A Review of the State of Current Stability Knowledge. International Conference on Ocean, Mechanical and Aerospace For Scientists and Engineer 2016: 277–290.

[10] Nebylov, A. V., Nebylov, V. A., Panferov, A. I., &Shepeta, A. P. Sea Plane Landing Control at Wave Disturbances. IFAC-PapersOnLine 2007; 6.

[11] IMO. Interim Guidelines for Wing-In-Ground (WIG) Craft. International Maritime Organization 2002; Pub. L. No. MSC/Circ.1054, 176. Worldwide.

[12] Rozhdestvensky, K. V. Wing-In-Ground Effect Vehicles. Progress in Aerospace Sciences 2006; 42: 211–283. doi:10.1016/j.paerosci.2006.10.001.

[13] Yun, L., Bliault, A., & Doo, J. Wings in Ground Effect. In WIG Craft and Ekranoplan: Ground Effect Craft Technology 2010; 1: 1–32. doi:10.1007/978-1-4419-0042-5.

[14] Pacificseaflights.com. What is WIG?. 2012; Retrieved December 19, 2018, from http://pacificseaflight.com/about-us-pacific-seaflight-wingship-wig/what-is-wing-and-

ground-effect-wig.

[15] Zagklis, P. Marketing Research for Greek. 2011; Retrieved December 19, Wingship Technology Corp 2018; from https://sites.google.com/site/hoverwingwigcraft/curriculum-vitae.

[16] IMO. Amendments to The Interim Guidelines for Wing-In-Ground (WIG) Craft. Worldwide: International Maritime Organization 2005; Pub. L. No. MSC/Circ.1054, 2.

[17] IMO. General Principles and Recommendations for Knowledge, Skills and Training for Officers on Wing-In-Ground (WIG) Craft Operating in Both Displacement and Ground Effect Modes. Worldwide: International Maritime Organization 2005; Pub. L. No. A.910(22), 15.

[18] Nebylov, A. Wing-In-Ground Flight Automatic Control Principles, Systems and Application Advantages. IFAC Proceedings 2001; 34(15): 542–547. doi:10.1016/S1474-6670(17)40784-1.

[19] Yun, L., Bliault, A., & Doo, J. WIG Craft Development. In WIG Craft and Ekranoplan: Ground Effect Craft Technology. Springer 2010; 1: 33–93. doi:10.1007/978-1-4419-0042-5.

[20] Wiriadidjaja, S., Mohamad, Z. H., MohdRafie, A. S., Elhadi, M., Fariduzzaman, & Hasim, F. Wing-In-Ground-Effect Craft as a Potential Domestic. In 34th AIAA Applied Aerodynamics Conference 2016: 1–11. doi:10.2514/6.2016-3879.

[21] Yin, C. M., Wiriadidjaja, S., Abd Majid, D. L., Romli, F. I., MohdRafie, A. S., &Zhahir, A. Review On the Cost and Performance of A WIGE Craft: A Commercialization Prospective. ARPN Journal of Engineering and Applied Sciences 2015; 10(21): 10027–10033. Retrieved from www.arpnjournals.com.

[22] Amir, M. A. U., Maimun, A., Mat, S., & Saad, M. R. Computational Analysis of Aerodynamic Characteristics for Wing in Ground Effect Craft in Lateral Stability. IOP Conf. Ser.: Mater. Sci. Eng. 2016; 152(012003): 9. doi:10.1088/1757-899X/152/1/012003.

[23] Staufenbiel, R. W., & Schlichtingt, U. Stability of Airplanes in Ground Effect. Journal of Aircraft. 1988; 25(4): 289–294. doi:10.2514/3.45562.

[24] Hahn, T., Drewelow, W., Dewitz, D., Kolewe, B., & Lampe, B. Analysis of Wing-in-Ground-Effect Vehicle with regard to Safety Ensuring Control. In IFAC Proceedings Volumes 2014; 47: 863–868. doi:10.3182/20140824-6-ZA-1003.01832.

[25] Yang, W., Yang, Z., & Collu, M. Longitudinal static stability requirements for wing in ground effect vehicle. International Journal of Naval Architecture and Ocean Engineering 2015; 7(2): 259–269. doi:10.1515/ijnaoe-2015-0018.

[26] Kornev, N., &Matveev, K. Complex Numerical Modelling of Dynamics and Crashes of Wing-In-Ground Vehicles. In 41st Aerospace Sciences Meeting and Exhibit, AIAA 2003; 41: 9. doi:AIAA 2003:600.

[27] Quah, J. Y. S. Stability, Performance & Control for a Wing in Ground Vehicle. 2005.

[28] Divitiis, N. De. Performance and Stability of a Winged Vehicle in Ground Effect. Journal of Aircraft 2005; 42(1): 148–157. doi:10.2514/1.4830.

[29] Nebylov, A., &Nebylov, V. WIG-craft Marine Landing Control at Rough Sea. In IFAC Proceedings Volumes 2008; 41: 1070–1075. doi:10.3182/20080706-5-KR-1001.00185.

[30] Daniel, D. Development of an Automatic Height Control System for WIG Crafts. In IFAC Proceedings Volumes 2008: 1088–1092. doi:10.3182/20080706-5-KR-1001.3176.

[31] Nebylov, A., Nebylov, V., & Fabre, P. Craft Flight Control Above the Waved Sea. IFAC-PapersOnLine 2015; 48(9): 102–107. doi:10.1016/j.ifacol.2015.08.067.

[32] Wang, H., Teo, C. J., Khoo, B. C., & Goh, C. J. Computational Aerodynamics and Flight Stability of Wing-In- Ground (WIG) Craft. Procedia Engineering 2013; 67: 15–24. doi:10.1016/j.proeng.2013.12.002.

[33] Kumar, P. E. Some Stability Problems of Ground Effect Wing Vehicles in Forward Motion. The Aeronautical Quarterly 1972: 41–52. doi:10.1017/S0001925900006302.

A CFD ANALYSIS OF WINGTIP DEVICES TO IMPROVE LIFT AND DRAG CHARACTERISTICS OF AIRCRAFT WING

Fatima Najeeb Khan[1], Beenish Batul[2],Ahmad Aizaz

Institute of Aviation Studies, University of Management and Technology, Lahore, Pakistan.

*Corresponding author e-mail: fatima.najeeb@umt.edu.pk

Abstract. The present study investigates the use of various wingtip devices to analyse the parameters of lift and drag for an aircraft wing. The coefficients of lift and drag are investigated in this research to optimize the wing design for enhancing the aircraft performance. A reduction in the drag produced due to wingtip vortices leads to reduced fuel consumption which contributes to the reduction in fuel emissions. The two-dimensional analysis is carried out for the selection of an apposite aerofoil by comparing the lift/drag characteristics of NACA 0012, 2415 and 23015 respectively at the velocity of 79.16 m/s at the angles of attack of 0°,4°,8°,12°,16° and 20°. The aerofoil section NACA 2415 is used to design the three-dimensional aircraft wing. For the analysis of the various wing tip devices the three-dimensional wing is incorporated with the spiroid winglet, blended winglet, wingtip fence and a mini-winglet. The CFD analysis for the wing designs is carried out for the take-off and landing phases of an aircraft's flight because the effect of vortices is the highest during these flight phases. The angles of attack range from 0° to 20°. The CFD results reveal that for the wing designs, the plain wing produced the highest drag and the blended winglet proved to be the wingtip device with the most beneficial design. The results obtained for the 30° cant angled blended winglet and 60° cant angled wingtip fence produces additional lift when compared to the results obtained for the counterpart designs. The results obtained from the analysis are in close correlation to the established use of the wingtip devices.

Keywords: Computational Fluid Dynamics, Wingtip Devices.

1. Introduction

The vortex theory of an airfoil's lift was developed in the early 1900's and most notably by Prandlt in 1918. The identification of the wing-tip vortices led to the initial explanation that the spinning mass of air was in the shape of an ellipse. It was Frederick W. Lanchester and Ludwig Prandtl's insightful notion and investigation of fluid mechanics which elaborated that the cylindrical air entraining a wing is equal in diameter to the span [1]. The creation of the vortices at the wingtip is defined as air's spillage from the wing's bottom side to the top surface, owed to the pressure difference between the top and the bottom surface. The production of drag caused by this difference in pressure above and below the wing's surface, induced drag, poses a great threat to the flight performance. The wingtip vortices cause the induced drag to increase hence causing a reduction in lift and a rise in drag. The vortex strength is dependent on the factors of weight, speed and shape of the wing of the aircraft. Owing to the dangerous impact of wake turbulence the take-off, climb and descent phases of flight are the most crucial flight phase. In an effort to achieve reduction in the production of induced drag a variation in the design of the wing tip was suggested as one of the possible techniques. Frederick W. Lanchester, in the year 1897 patented the use of end-plates at the wingtips for monitoring the vortices produced at wingtips [2]. In the early 1970's NASA's Richard T. Whitcomb invented wingtip extensions which were almost vertical with the intention to enhancing the wing lift-to-drag performance by producing an increase in the parameters [3]. Winglets effectually avert the flow of air from the tip vortex in turn reducing the drag.

In this research CFD study is performed to investigate the effect of wingtip devices on the coefficients of lift and drag. The values attained are analysed to suggest the most optimum design during take-off and landing. This study also analyses the effect the varying cant angle produces on the flight characteristics of the aircrafts. Earlier research has been conducted for blended winglets [4], multiple winglets [5] and spiroid winglets [6] however this research specifically combines the results for the not only the wingtip devices being used but also presents the effects produced by changing the cant angles of the selected devices.

2. Literature Review

During the take-off phase of flight for an aeroplane the formation of wingtip vortices contributes up to 60% of the total drag [7]. Raj and Thomas [8] concluded that for the period of cruise 40% of the total drag and during take-off conditions 80-90% of the over-all drag is owing to induced drag. Mattos, Macedo, and Filho [9] further elaborated that with the addition of wingtip devices to the wings aim at the decrease in the production of induced drag. The induced drag makes up to 30% to 40% of the entire drag for a transport category aircraft at the cruise condition for a long-range flight and also has a noticeable downgrade effect on the climb performance for an aircraft. In 1986 NASA [10] assessed various drag reduction devices at the wing tip; winglets, feathers, and sails, and observed that the use of such devices designed as an integral part of the wing could lead to the decrease in drag created due to lift producing components by 10% to 15%. Analogous research was also carried out by Smith et al. [11] to examine the decrease of induced drag by investigating the capability of multi-winglets for the same aircraft wing's span.

In 1994, Louis B. Gratzer acquired the patents for the most widely used winglet, blended winglet. Gratzer [12] defined the shape of blended winglets as a wing-like surface encompassing a blended surface for enhancing lift attached to an aircraft's wingtip for a selected surface size of the wing to attain minimum induced drag. Aviation Partners Inc. was the first to introduce blended winglets for the Gulfstream aircraft in 1992. For the enhancement in the design of blended winglets Aviation Partners Inc. and Boeing collaborated in the year 1999. The finalisation of the design and patent of the blended winglet lead to its installation on a number of Boeing aircrafts and its variants i.e. B737, B757, and B767. The used of these winglets contributed to superior take off performance, reduction in engine maintenance costs and also contribute in lowering noise and emissions along with the 6% reduction in drag. Narayana et al.'s [13] aerodynamic analysis for a rectangular wing using XFLR5 deduced that a 4.33% increase in the C_L/C_D ratio for a wing was recorded for an increase in wing span, for a wing using winglets the calculated increase is 6.63% and 4.83% increase was evident for a wing with blended winglet. The evaluation of five types of wingtip devices for commercial aircraft model for cruise conditions at various angles of attack was investigated by Gavrilović et al. [14]. The analysis included spiroid wingtip, blended winglet, maxi wingtip, wing fence and spiroid 2. The results obtained showed the maximum percentage improvement of 5% due to the use of maxi wingtip and 3.78% for blended winglet. Another CFD simulation emphasised on the wing of a Boeing 737-800 with and without winglet was conducted by Atique et al. [15]. The analysis determined that at an angle of attack of 4° the drag decline was about 40.13% at 0.35 mach in comparison to the lessening of drag for flight between 0.5 Mach number and 0.6 Mach number was a minimum of 3.47%.

3. Methodology

3.1. Design of 2D aerofoil and 3D wing

In order to investigate the usage of wingtip devices, the selection of the 2D aerofoil to be used for the construction of the 3D is performed. The 2D aerofoil sections for NACA 0012, NACA 2415 and NACA 23015 have been designed using a CSV file for importing coordinates to make the profile for the selected aerofoils. The 3D wing is constructed by using the CSV coordinate file format in the 3D CAD module of STAR CCM+. The values for the wing design have been postulated by closely following Airbus A320's wing. The wing span is measured as 15 m. The wing has also been designed with the wingtip devices; blended winglet, wingtip fence, spiroid winglet and mini-winglet for the study. The complete wing design is shown in Fig. 1.

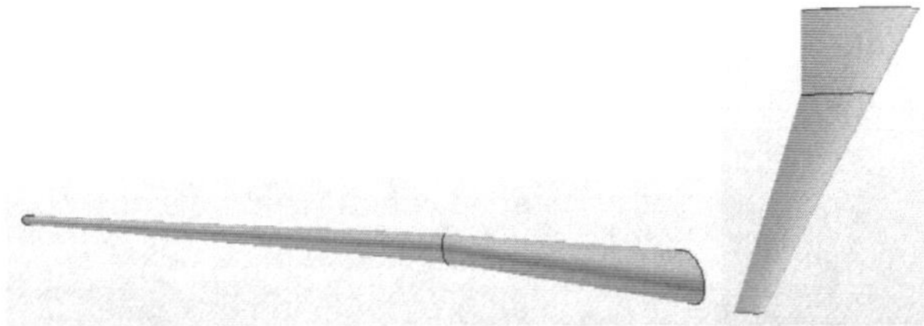

Fig. 1: 3D wing completed design

The basic configuration for the plain wing with an augmented aerofoil profile at the tip of the wing is created for the blended type winglet. The design of the winglet consists of varying cant angles of 30°, 45° and 60° to evaluate the optimum design characteristic. The layouts of the varied cant angled winglets are shown in Fig. 2.

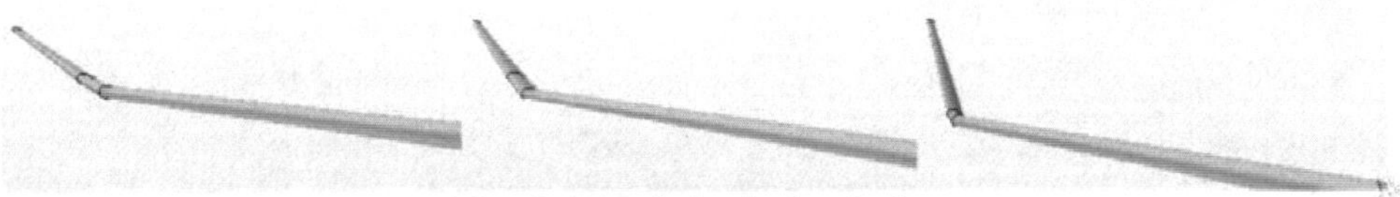

Fig. 2: Blended winglet design

The spiroid winglet is designed with a simplified semi-circular profile, the second design is based on a varying aerofoil chord length and the third design incorporated parallelogram shaped spiroid. The three configurations of the spiroid designed winglets are shown in Fig. 3.

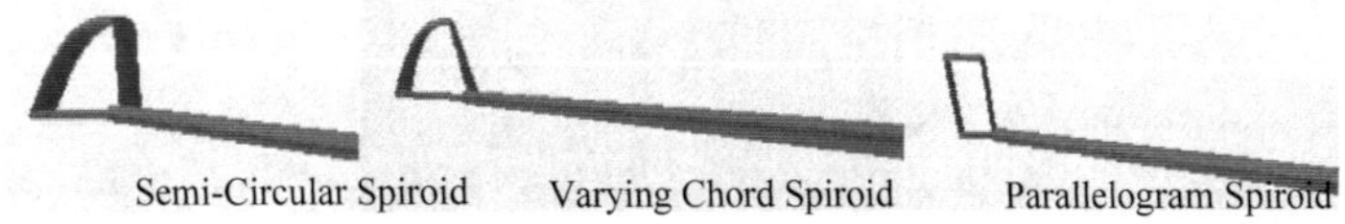

Fig. 3: Spiroid winglet designs

The profile of the wingtip fence incorporates an extended surface above and the other below the wingtip. NACA 2415 is used to form the wingtip fence. The design is further constructed by varying the aerofoil's upper surface cant angle. The cant angles at 90° and 60° are used for the design of the wingtip fence for the upper surface. Fig. 4 shows the designs of the main wing with the wingtip fence appended.

IOP Conf. Series: Materials Science and Engineering **642** (2019) 012006 doi:10.1088/1757-899X/642/1/012006

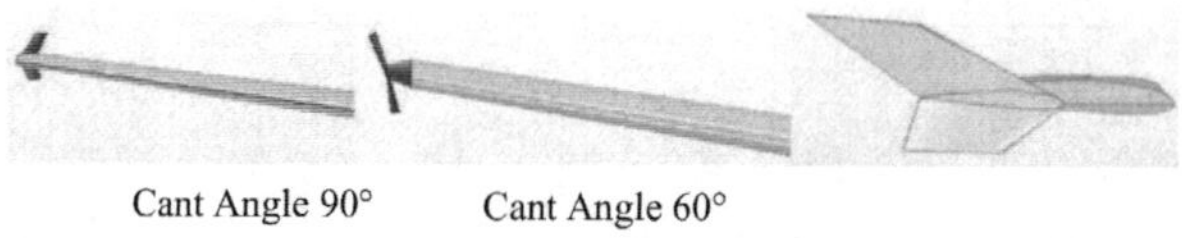

Fig. 4: Wingtip fence designs

The tenacity for constructing the mini-winglet is to review the influences that it would produce on the flight performance parameters since the size of the winglet is reduced in comparison to similar winglets i.e. blended winglet. A 60°cant angle has been specified for the mini-winglet as shown in Fig. 5.

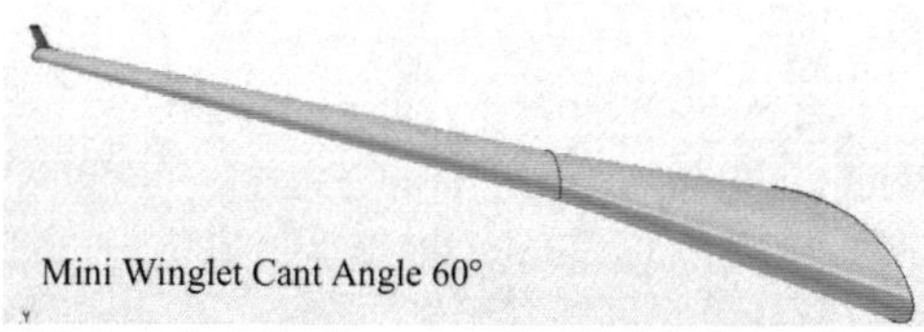

Fig. 5: Main wing design attached with the Mini-Winglet

3.2. Computational Domain and Boundary Conditions

A far-field boundary is bullet shaped as shown in Fig. 6. is shaped for simulating the air flow for the three 2D aerofoils. The computational domain's inlet and the walls have been sketched to make sure that the flow fully develops past the aerofoil resulting in not any errors caused by reverse flow. For the 2D aerofoil analysis the walls of the computational domain are specified suiting the specification of the simulation. A 'bullet-shaped' domain is constructed for the 3D wing analysis similar to the 2D domain. The domain shape and the wing are shown in Fig. 6. The domain size has been specified by using the span of the wing as the reference length.

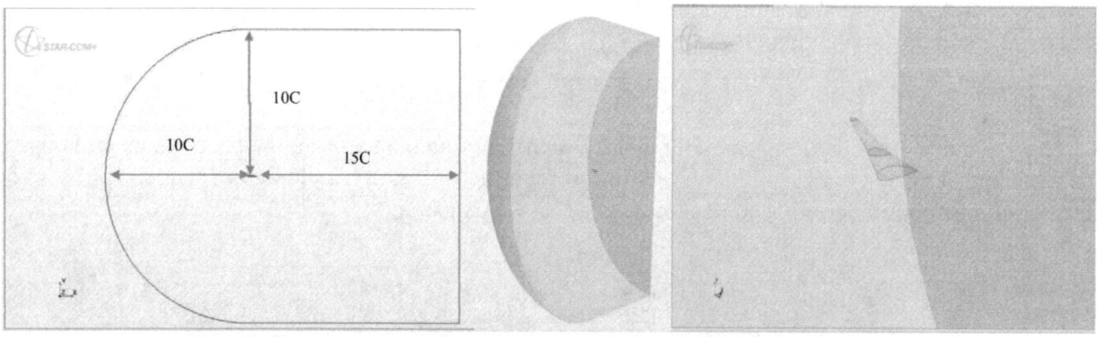

Fig. 6: Computational domain for aerofoil and 3D wing simulation

3.3. Mesh Generation

Volumetric control blocks of the shape of block and cylinder have been made for the 2D analysis of aerofoils. The mesh has been refined around the aerofoil and further refinement has been done around leading and trailing edge of the aerofoil while the rest of the computational domain's mesh is coarse. Further the values for the prism layer stretching the prism layer thickness, number of prism layers, base size have been specified. The computational domain's mesh is shown in Fig. 7.

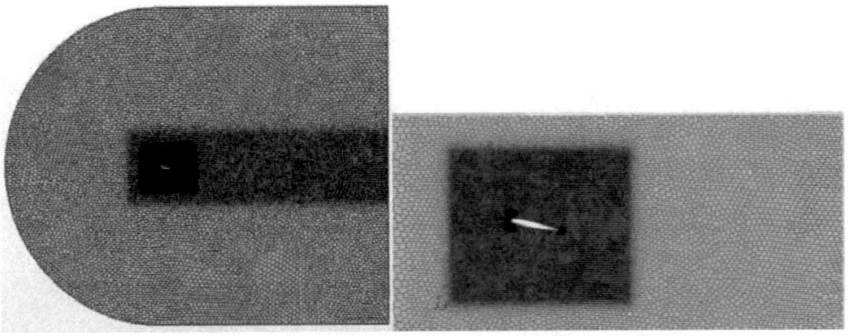

Fig. 7: Mesh of 2D domain and aerofoil

The 3D computational domain's meshing is done keeping in consideration the limitation of the resources available for computations as well as time. Three different mesh refinement options have been used for achieving mesh refinement. The approximate cell size measurement on the wing surface has formed the base size. The largest distance on the surface of the computational domain on the 'Inlet' upstream region is specified as the maximum cell size. The mesh generated for the computational domain and the 3D wing with blended winglet are shown in Fig. 8 below.

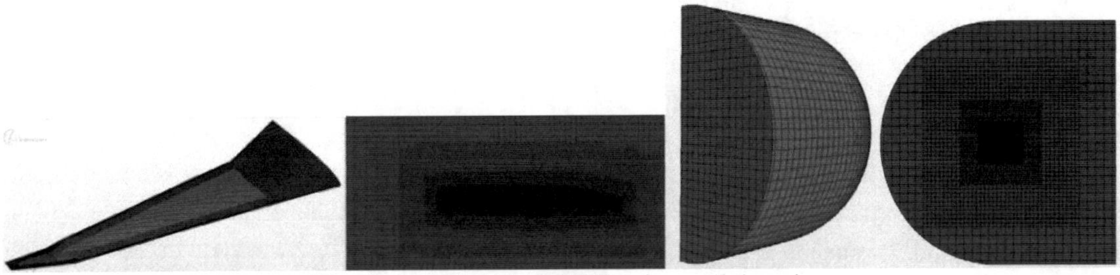

Fig. 8: Completed mesh 3D wing and domain

3.4. Simulation Parameters

The Reynolds number calculated for the simulations is 7.8×10^6. For the 2D analysis the turbulence model k-ε is selected. The Realizable Two-Layer K-Epsilon model supported with the two layers all y+ wall treatment is also selected as it combines the Realizable K-Epsilon model with the two-layer approach. The values for initial velocity, dynamic viscosity for air and density for both the 2D and 3D models is presented in Table 1. The reference area specified for the two-dimensional aerofoil is 1.5 m^2. The turbulence model selected for the three-dimensional wing analysis is Spalart-Allmaras model.

Table 1: Initial conditions used for 3D simulations

Parameters	**Take-off**	**Landing**
Initial Velocity	79.16 m/s	68.89 m/s
Dynamic Viscosity	1.826 x 10^{-5} N.s/m^2	1.826 x 10^{-5} N.s/m^2
Density	1.2 kg/m^3	1.2 kg/m^3

3.5. Convergence Criteria and Mesh Independence Check

For the fulfilment of the convergence criteria to validate the simulations the magnitude of the simulation residuals should reduce by two or three orders as a first check for the convergence criteria. Fig. 9 shows the converged solution obtained for both the 2D aerofoil and 3D wing designs.

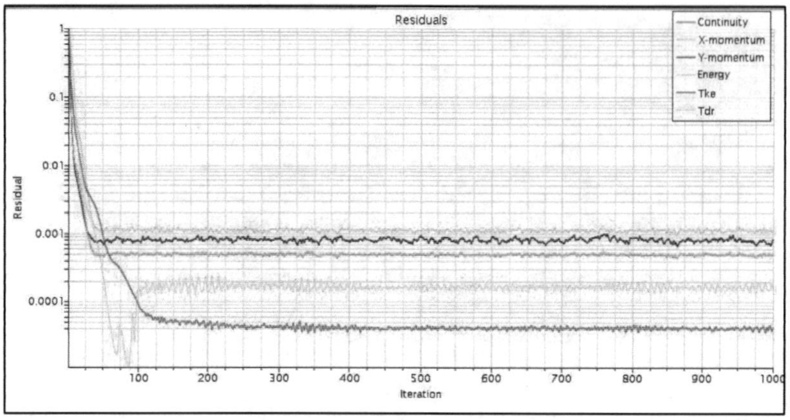

Fig. 9: Converged residuals

The use of higher mesh elements for refinement carried out for the 3D wing model confirms the mesh independence for the final model. The percentage difference of 0.13% is calculated between M2 and M3 while a larger difference is obtained between the coarse mesh, M1 and finer mesh M2 as is seen in Fig. 10. From the results obtained it is established that the values used for M2 will be used to mesh all the models for simulations.

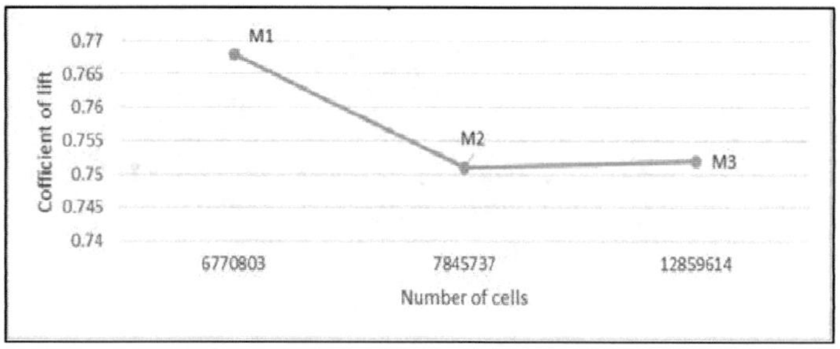

Fig. 10: Mesh Independence check

4. Results and Discussion

4.1. Aerofoil comparison results

The lift and drag coefficients obtained for the three NACA profiles have been presented in Fig. 11.

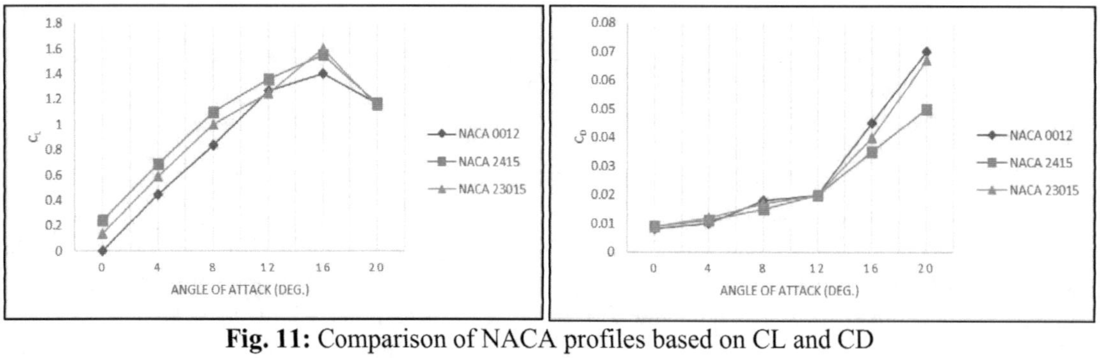

Fig. 11: Comparison of NACA profiles based on CL and CD

From the values obtained for the 2D analysis, it can be seen that the value of C_L for NACA 2415 is higher as compared to the two aerofoils. Even though similar stalling characteristics can be observed for the three aerofoils, the lift coefficient for NACA 2415 is greater at higher angles of attack compared to the NACA 0012 and NACA 23015 aerofoils. Similarly, the value of coefficient of drag is lower for NACA 2415 in comparison to the other studied aerofoil designs. Hence NACA 2415 is used for the construction of the rest of the 3D wings.

4.2. Results comparison for take-off and landing conditions

The comparison of the results obtained for the value of C_L and C_D for plain wing and the wings with wingtip devices is shown in Fig. 12. The design with the best lift characteristics has been presented for the comparison.

The analysis of the coefficient of lift values shows that during take-off the blended winglet provides the highest value of C_L in the range of 0° to 20°. The spiroid parallelogram winglet also provides more lift and stalls later than the plain wing. The least lift increase in comparison with the plain wing is produced by mini winglet. From the results obtained it can thus be deduced that the blended winglet provides the best lift due to the reduction in wingtip vortices. The values obtained for coefficient of drag exhibits similar increase in the value for all the wing designs. However, the value of drag at higher angles of attack is smaller than the others for the parallelogram spiriod winglet.

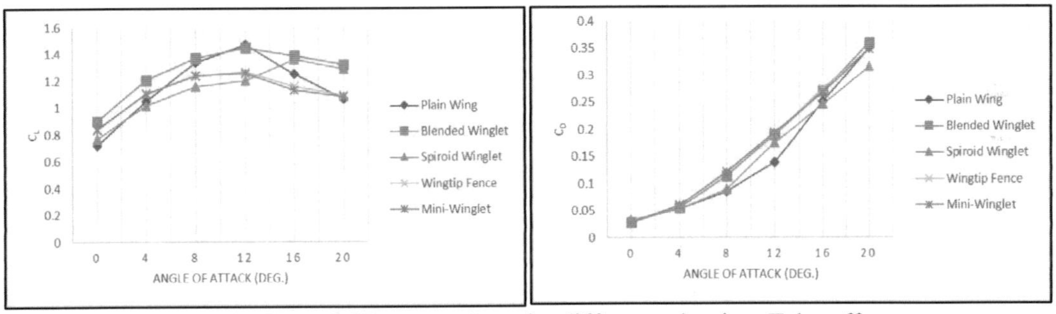

Fig. 12: CL and CD comparison for different wingtips- Take-off

The results of the wingtip devices for the landing phase conditions are presented in Fig. 13.

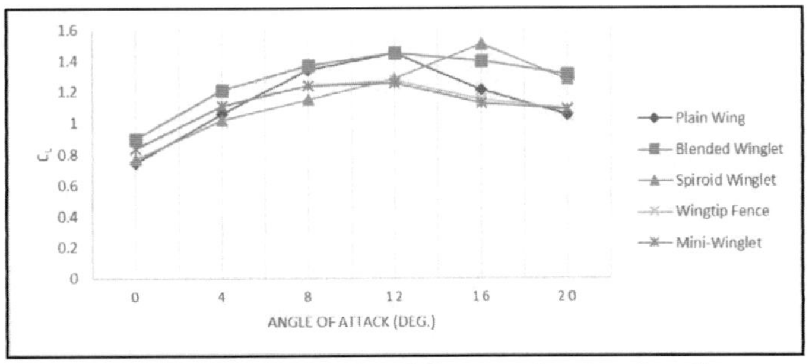

Fig. 13: CL comparison for different wingtip - Landing

The simulations run for the flight phase of landing yields the results that show the blended winglet produces the most improvement for the C_L values. The winglet shows a steady increase in the values of lift coefficient that is obtained for the selected range of angles of attack. The critical angle of attack is same as that observed for the plain wing design that is 12°. However, at 0° angle of attack, the maximum increase in lift is obtained for the wing. The percentage increase in lift attained at higher angles of attack reveals an increase of about 15% at the angle of attack of 16° and 24% at the angle of attack of 20°. The wingtip devices which produce similar performance parameter output are the wingtip fence, spiroid winglet and the mini-winglet. At 0° angle of attack the increase in lift is for the spiroid winglet is the least with only an increase of 3% while an increase of 12% is calculated for the wingtip fence and mini-winglet. As the angle of attack is further increased the values of C_L acquired decrease. The results for the wingtip fence and mini-winglet show a decrease in lift coefficient for the angles of attack 8°, 12° and 16°. The decrease in C_L is calculated at about 8%, 12% and 5% approximately for the two wingtip devices respectively. The results for the spiroid winglet also show a reduction in the obtained values; however, the values begin to decrease as soon as the angle of attack increases from 0°.

The results for the values of C_D are shown in Fig. 14. The maximum percentage decrease at 0° angle of attack for the values of C_D has been attained for the blended winglet design. The results show a 22% decline in the value of C_D for the blended winglet when compared with the plain wing. For the same angle of attack the spiroid winglet produced only 2.81% decrease for C_D. In comparison to the plain wing as the angle of attack increases to 4° the drag also increases. For the blended winglet and spiroid winglet, the decrease is 1.2% and 0.26% respectively. An increase in the drag is observed for all the other simulated wingtip devices. For all the wingtip types, the analysis shows that with the increase in the angle of attack a significant increase in the drag coefficient occurs. The highest increase of about 40% is obtained for the wingtip fence. For 12° and 16° angle of attack the same trend continues. The spiroid winglet at 16° angle of attack is the only wingtip device that shows a decline in the value of C_D. At 20° angle of attack decrease in drag is about 9% which is obtained by using the sprioid winglet and 1.2% reduction in C_D is obtained for the blended winglet.

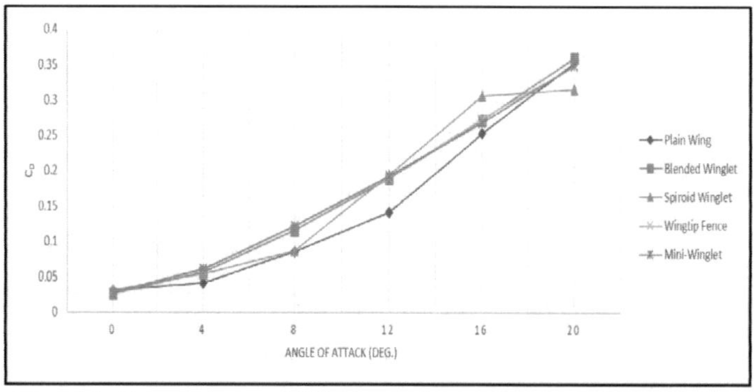

Fig. 14: Comparison of CD for various wingtips – Landing

The results obtained in this research are in close correlation to the study conducted by Beechook and Wang [13]. In their study, a comparison is done of the results obtained by conducting a wind tunnel experiment and CFD analysis for a plain wing and a wing with blended winglet is [13]. The conclusion stated that the increase in lift obtained for the blended winglet at 45° cant angle is greater than the lift obtained for blended winglet at 60° cant angle. The result analyses for both the researches show that the trend of increase in lift coefficient and decrease in drag coefficient obtained are identical. The results obtained in this research for the plain wing and blended winglet show the same results with the increase in lift being greater for the winglet with 45° cant angle in comparison with winglet having 60° cant angle and plain wing.

5. Conclusion

The effect of using wingtip devices on wings has been analysed in this CFD study by analysing the values of lift and drag coefficient. The decline in the drag coefficient and increment in the lift reveals the significance of the various wingtip devices. The comparison of coefficients of lift and drag attained from the initial investigation revealed the aerofoil NACA 2415 to be the most suitable aerofoil for the modelling of the wing's three-dimensional design. The three-dimensional plain wing design without and equipped with devices of wingtips is modelled and designed for two flight stages; take-off and landing. Various cant angles for the selected wingtip devices, blended winglet, spiroid winglet and the wingtip fence, were varied for the analysis. The comparison and analysis of the results obtained concludes that the 30° cant angle blended winglet and the 60° cant angle wingtip fence when compared to their counterpart designs produced more lift. The comparison of the various spiroid

IOP Conf. Series: Materials Science and Engineering **642** (2019) 012006 doi:10.1088/1757-899X/642/1/012006

winglet designs reveal that the design configuration of the spiroid winglet with parallelogram produced less drag and more lift. Hence the blended winglet with 30° cant angle can be regarded as the most suitable winglet as it produces the highest rise in lift coefficient and a comparable decrease in drag coefficient.

6. Conclusion

The authors express their gratitude to Dr. Liz Byrne, University of Hertfordshire, for the supervision and support extended by her for this work.

References

[1] Stinton D. The anatomy of the aeroplane. Oxford: Blackwell Science; 1997.
[2] Chambers J. Concept to reality. National Aeronautics and Space Administration; 2000.Washington, D.C.
[3] Beechook A, Wang J. Aerodynamic analysis of variable cant angle winglets for improved aircraft performance. In2013 19th International Conference on Automation and Computing 2013:1-6.
[4] Anderson Jr JD. Fundamentals of aerodynamics. Tata McGraw-Hill Education 2010.
[5] Smith M, Komerath N, Ames R, Wong O, Pearson J. Performance analysis of a wing with multiple winglets. In19th AIAA Applied Aerodynamics Conference 2001: 2407. doi: 10.2514/6.2001-2407.
[6] Shelton A, Tomar A, Prasad JV, Smith M, Komerath N. Active multiple winglets for improved UAV performance. In22nd Applied Aerodynamics Conference and Exhibit 2004; 43(1): 110-116. doi: 10.2514/1.13987.
[7] Raj WG, Thomas TA. Design and analysis of spiroid winglet. International Journal of Innovative Research in Science, Engineering and Technology 2015;4(3). doi: 10.15680/IJIRSET.2015.0403070.
[8] Mattos B, Macedo A, Silva Filho D. Considerations about winglet design. In21st AIAA Applied Aerodynamics Conference 2003: 3502. doi: 10.2514/6.2003-3502.
[9] Yates JE, Donald CD. A fundamental study of drag and an assessment of conventional drag-due-to-lift reduction devices. NASA Contractor Report 4004 1986.
[10] Smith M, Komerath N, Ames R, Wong O, Pearson J. Performance analysis of a wing with multiple winglets. In19th AIAA Applied Aerodynamics Conference 2001: 2407-2417. doi: 10.2514/6.2001-2407.
[11] Gratzer LB, inventor; Gratzer Louis B, assignee. Blended winglet. United States patent US 5,348,253. 1994 Sep 20.
[12] Aswatha Narayana PA. Design and analysis of wingtip devices. History 2015;31(139):48-55.
[13] Gavrilović NN, Rašuo BP, Dulikravich GS, Parezanović VB. Commercial aircraft performance improvement using winglets. FME Transactions 2015;43(1):1-8.
[14] Atique MS, Salam MA, Nafi AS, Probha NN, Barman S. Aerodynamics of Winglet: a Computational Fluid Dynamics Study. Global Science and Technology Journal 2015; 3: 10-24.

Experimental Analysis of Coir Fiber Sheet Reinforced Epoxy Resin Composite

Van Quy HO, Song Thanh Thao NGUYEN

Department of Aerospace Engineering, Ho Chi Minh City University of Technology, Vietnam

*Corresponding author e-mail: 1413206@hcmut.edu.vn

Abstract. In recent years, the industrial demand for composite materials is increasing rapidly. However, most composite materials are produced from synthetic fibers and resins. As a result, the current production of composite materials still has many problems and the most worrying concern is the environment and human health. From those challenges, natural fiber composite is gradually becoming a worldwide research and development trend. Natural fibers are an abundant and renewable resource, so their cost is relatively low compared with other conventional fibers. They are eco-friendly, biodegradable and reduce the problem of solid waste production when used to replace non-degradable fillers. The objective of this study is to approach and learn about natural fiber composite, which is divided into two main stages. In the first one, we learn about natural fibers and matrixes of natural fiber composites. In the second stage, material testing experiments were performed on the coir fiber sheet reinforced epoxy resin composites to determine their mechanical properties as well as to find the mechanical properties of the coir fiber sheet using the rule of mixtures theory. The experimental results show that the non-split coir fiber sheet has the similar properties with ones of single coir fiber and results in a higher mechanical performance compared to the split coir fiber sheet. Additionally, the results show a medium stiffness, small density and low-cost composite material. Therefore, the further purpose of this study is to apply natural fiber composites in aeronautical engineering such as manufacturing small unmanned aerial vehicle.

Keywords: Natural fiber composite, Polymer matrix, Coir fiber, Tensile test, Rule of mixtures.

1. Introduction

A composite material is a material made from two or more different component materials in order to make use of the advantages of components to produce the material having superior properties compared to individual components. About history, simple natural composite fiber materials exist from a long time ago. It can be found in wood in which cellulose fibers are held together by lignin. The connection of cellulose is responsible for adequate strength and density in plants and trees.

Nowadays, composite material industry has been developing strongly all over the world with many remarkable achievements and new inventions. However, in addition to development, many negative issues have risen, specifically the negative impact on the environment and human health. Consequently, research and development of natural fiber composite (NFC) are of interest. With many valuable features and advantages of NFC such as low specific gravity, low cost in production process, relatively high in strength and stiffness and especially eco-friendly and biodegradable [1-2], which is less harmful to both human and environment, the application of natural fibers for reinforcement in composite materials has increased significantly in the past few years.

Natural fibers include all those derived from plants, animals and mineral resources [3]. In the scope of this study, we focus on studying and understanding the mechanical properties of natural plant-derived fibers. Natural fibers derived from plants are mainly composed of cellulose such as cotton, jute, flax, hemp, sugarcane, bamboo and many other plant fibers. These fibers are taken from different parts of the plant such as fibers from the seeds (cotton fibers), the leaves (sisal fibers), the bark (flax, hemp, jute, rattan or even banana and sugarcane fibers), the fruits (coconut fibers from coconut husks) and fibers from stems (bamboo, grass, wood fibers) [3]. In those mentioned fibers above, the ones are taken from the bark and stem will have the best mechanical properties and have great potential in the process of making reinforced fibers from natural fiber composites (NFCs).

2. Natural vegetable fibers

In this study, the authors introduce and learn about the mechanical properties of natural fibers that can be exploited in Vietnam.

2.1. Cotton fiber

Cotton fibers are soft and evenly spun; fibers are taken from the seeds of cotton plants - a shrub that normally grows in tropical and subtropical regions. In Vietnam, cotton trees are mostly grown in the Central Highlands and Southeast. Cotton fibers are friendly to human skin and hypoallergenic, which makes cotton fibers become an important ingredient in the textile industry. In addition, with relatively good mechanical properties, cotton fibers are also used today as reinforcing components in composite materials.

2.2. Jute fiber

Jute is a long and soft plant fiber that can be pulled into coarse, hard yarn. The fibers are relatively white to slightly brown and are 1 to 4 meters long. Some advantages of jute yarn are: it is 100% biodegradable and recyclable, so it is very friendly to the environment; it has high tensile strength, low elongation. In Vietnam, jute trees are grown in the Red River Delta region and Vietnam is one of the top ten jute producing countries according to the 2014 report. Just like cotton, jute yarns are mainly used in the garment and packaging industry. Besides, thanks to the available mechanical properties, jute fiber is also now studied and applied in the composite material industry.

2.3. Banana fiber

Banana fibers, which are derived from banana bark, have the following characteristics: it is a very strong fiber; it is similar to bamboo fiber but better fineness and smaller elongation; it is lightweight; it absorbs strong moisture but releases moisture very quickly; it can biodegrade. In Vietnam, banana trees are grown in many areas and it can be said that they are easy to grow and exploit. In previous years, the application of banana fiber is very limited and is mainly used to manufacture items such as ropes and carpets. However, recently, after studying the properties of banana fiber, with good mechanical properties and its usability, it is also used to make reinforced fibers in NFCs.

2.4. Bamboo fiber

Bamboo fibers are mainly cellulose fibers, extracted or made from natural bamboo. Bamboo has higher compressive strength than wood, brick or concrete and tensile strength can be compared with steel. Bamboo fibers come in different forms such as hardened strips produced by cutting bamboo stems into thin strips; rayon, also known as semi-synthetic fibers, are created by regenerating cellulose extracted from bamboo; Smooth fiber is produced by a machine-assisted process. Bamboo has a great economic and cultural significance in South Asia, Southeast Asia, East Asia and especially for Vietnam. Bamboo has many uses such as manufacturing food, building materials, household items, agricultural tools, etc. Today, with valuable mechanical properties, bamboo fibers are being researched to have applicable in NFCs.

2.5. Coir fiber

Coconut fiber, also known as coir, is a natural fiber extracted from the part between the hard-inner crust and the outer layer of the coconut. There are two types of coconut fiber: brown coconut fiber extracted from the old coconut husk; White coconut fiber is extracted from young coconut shells. The brown coir fiber is strong and highly abrasion resistant. The white coir fiber is smoother, but also weaker. Since ancient times, coir fiber has been used by humans for many different purposes such as making clothes, making ropes, making carpet coverings, etc. Today, with available mechanical properties, coir fiber is also researched and applied in the natural fiber composite industry.

Table 1 shows that the mechanical properties of coir fiber are lower than other natural fibers that can be exploited in Vietnam. However, because of recent available market reasons in Ho Chi Minh City, coir fiber in the form of the sheet is chosen to fabricate composite material on which perform the experiments.

Table 1: The mechanical properties of coir fiber compared to other natural fibers [4-7]

Fiber	Coir	Cotton	Jute	Banana	Bamboo
Density (g/cm^3)	1.2	1.5 – 1.6	1.3 – 1.5	1.35	0.6 – 1.1
Length (mm)	20 – 150	10 – 60	1.5 – 120	150	100
Diameter (μm)	100 – 450	–	40 – 350	50 – 250	240 – 330
Tensile strength (MPa)	131 – 503	287 – 800	393 – 800	600	503
Young's modulus (GPa)	4 – 6	5.5 – 12.6	10 – 55	17.85	35.91
Elongation at break (%)	15 – 30	7 – 8	1.5 – 1.8	3.36	1.4

3. Polymer matrices in natural fiber composite

3.1. Polyethylene

Polyethylene (PE) is a thermoplastic polymer with a variable crystal structure and an extremely large range of applications. Polyethylene (PE) is often classified into many major compounds, most commonly Low-Density Polyethylene (LDPE), Linear Low-Density Polyethylene (LLDPE), High-Density Polyethylene (HDPE) and Ultra High Molecular Weight Polypropylene (UHMWPE). Among them, LDPE and HDPE, which have applicable mechanical properties, are usually used a lot in making matrices for NFCs.

3.2. Polypropylene

Polypropylene (PP) is a thermoplastic polymer produced via chain-growth polymerization from propylene monomer. Its properties are similar to polyethylene, but it is slightly harder and better resistant to heat. It is a material having good mechanical properties and high chemical resistance. Thanks to suitable mechanical properties, polypropylene is also commonly used as a matrix in NFCs.

3.3. Phenol formaldehyde resin
Phenol-formaldehyde resins, also known as phenolic resins, are some synthetic resins created by producing a reaction between phenol and formaldehyde. Thanks to appropriate mechanical properties, phenol formaldehyde resin has also been studied for application as a matrix in NFCs.

3.4. Epoxy resin
Epoxy resins, also known as polyepoxides, which have favorable mechanical properties, high heat and chemical resistance along with good strength and adhesion, are used for many different purposes. Applications of epoxy resins are very wide, such as coatings, adhesives, electronic applications, and especially it is used in aerospace industry as a matrix for composite materials with the reinforcement of fibers like glass fiber, carbon fiber and Kevlar fiber. In addition, it is now also used for research and application as a matrix in NFCs.

Table 2 shows mechanical properties of common matrix materials in NFCs.

Table 2: The mechanical properties of epoxy resin compared to other polymer matrices [8-10]

Polymer matrix	LDPE	HDPE	PP	Phenolic resin	Epoxy resin
Density (g/cm^3)	0.91 – 0.93	0.94 – 0.96	0.9 – 0.92	1.21	1.1 – 1.4
Tensile strength (MPa)	5 – 25	15 – 40	25 – 40	45	35 – 100
Young's modulus (GPa)	0.1 – 0.3	0.5 – 1.3	0.9 – 2	6.5	3 - 6
Elongation at break (%)	400	150	80	1.2	1 – 6
Strain at yield (%)	19	15	–	–	–
Poisson's ratio	–	0.46	0.43	–	0.3

4. Experiments

4.1. Experimental system
Experiments are performed on the miniature testing machine, which is designed and manufactured by our own research team as shown in Fig. 1. In this testing machine, a loadcell is responsible for data acquisition of load and a strain gauge takes data of longitudinal strain during testing. Thanks to the data from the loadcell and the strain gauge, the stress-strain curve is defined, from which we can determine the values of tensile strength and Young's modulus. Because the test system has just two data acquisition channels, the transverse strain is not be measured. So, the system also incorporates another strain measurement method based on the Digital Image Correlation (DIC). The results from strain measurement by DIC allows to define the Poisson's ratio and then the shear modulus. However, the accuracy of this method depends on the resolution of image.

Fig. 1: The system used to test mechanical properties of composites

4.2. Experimental standard

Currently, for the tensile testing of composite materials, it is common to use standard ASTM D3039. However, for some other composite materials, the proposed standard is ASTM D638. In summary, ASTM D3039 is recommended for highly oriented and/or high tensile modulus fiber reinforced polymer composites. ASTM D638 is recommended for randomly oriented, discontinuous, moldable, or low reinforcement volume composites or composites with tensile modulus less than 20000 MPa. From those factors, for the coir fiber sheet reinforced epoxy resin composites, the standard chosen for tensile testing is ASTM D638. All the tests are conducted under displacement control at loading rate of 5 mm/min.

4.3. Specimens

In this study, coir fiber was selected in the form of a sheet (coir tape) with a width of 20 cm and a thickness of 1-2 mm (Fig. 2). There is no given specification for this coir tape.

Fig. 2: The coir fiber sheet (coir fiber tape)

The YDL 586 type epoxy resin is used with the specifications provided by the company including:

- Young's Modulus: 2.8 - 3.4 GPa
- Tensile strength: 60 - 70 MPa
- Elongation at break: 4 - 7%

The coir/epoxy composite material sheets are fabricated by the compression molding method. The dumbbell specimens are machined out from these sheets with the dimensions following the standard ASTM D638 - Type I as shown in Fig. 3. The width and length of the gauge length is 13 mm and 57 mm respectively.

After that, the strain gauge is stuck on surface of specimen. In addition, to be able to measure simultaneously the strain by DIC, four points are also marked on surface of specimen at the effective position (the middle of specimens) with the distance of about 10mm apart as shown in Fig. 4.

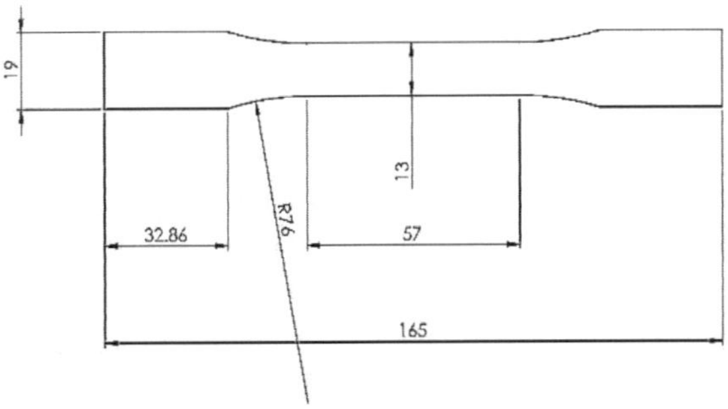

Fig. 3: The dimension of specimens according to ASTM D638 - Type I

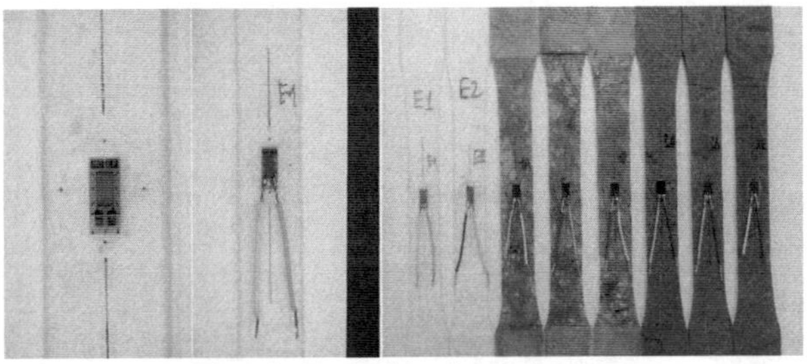

Fig. 4: Completed specimens

4.4. Digital Image Correlation

In DIC method, strains are computed from the displacement of four marks laid at the surface of the specimen.

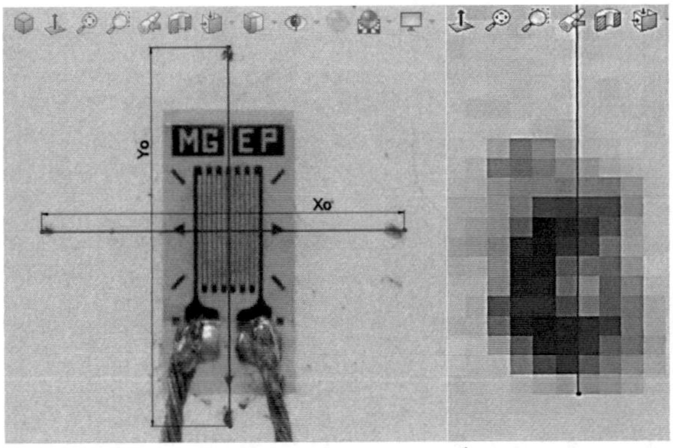

Fig. 5: DIC method: a) position of four markers, b) image of one marker displayed in pixels in Solidworks software

The longitudinal and transverse strains are calculated following Eq. (1) and (2),

$$\varepsilon_1 = \frac{y_i - y_o}{y_o} \quad (1)$$

$$\varepsilon_2 = \frac{x_i - x_o}{x_o} \quad (2)$$

where y_i and y_0, x_i and x_0 respectively stand for the current and initial distance between longitudinal and transverse marks.

Thus, the Poisson's ratio is defined by Eq. (3),

$$\nu_{12} = -\frac{\varepsilon_2}{\varepsilon_1} \quad (3)$$

4.5. Rule of mixtures

Rule of Mixtures (RoMs) is a method of approach to approximate estimation of composite material properties, based on an assumption that a composite property is the volume weighted average of the phases (matrix and dispersed phase) properties. It is the simplest method to determine the elastic properties for a unidirectional composite material. The Young's modulus of the matrix and fiber are taken as E_m and E_f, similarly Poisson's ratios are taken as ν_m and ν_f. The quantities of two components inside composite in terms of volume fraction is,

Longitudinal properties,

$$E_1 = E_f V_f + E_m V_m \quad (4)$$

$$\upsilon_{12} = \upsilon_f V_f + \upsilon_m V_m \quad (5)$$

Where V_f and V_m are volume fractions.

Transverse properties,

$$E_2 = \frac{E_m E_f}{E_m V_f + E_f V_m} \quad (6)$$

$$\upsilon_{21} = \frac{\upsilon_{12}}{E_1} E_2 \quad (7)$$

Shear properties,

$$G_{12} = \frac{G_m G_f}{G_m V_f + G_f V_m} \quad (8)$$

Where $G_f = \dfrac{E_f}{2(1+\upsilon_f)}$ (9)

$$G_m = \frac{E_m}{2(1+\upsilon_m)} \quad (10)$$

RoMs also allows to determine inversely the mechanicals properties of component materials when the properties of composite are given.

5. Experimental results

The goal of study is not only to determine the mechanical properties of composite materials but also to determine the mechanical properties of components including coir fiber and epoxy resin. In addition, it is also important to study the effect of the distribution and percentage of fiber volume on the mechanical properties of the composite material. With the current experimental conditions, it is not possible to measure the mechanical properties of single coir fiber, so the properties of the fiber are attributed to the properties of the coconut sheet by theoretical calculation. To solve these requirements, the experiments are performed on three specimen groups including epoxy resin group and two composite groups.

5.1. Density and volume fraction

To determine the mechanical properties of composite materials, it is required the density of the component materials as well as the percentage of fiber volume.

The determination of density of epoxy resin and coir fiber is carried out according to the following steps:

- Step 1: Determine the density of epoxy resin using electronic scales and volumetric flasks.
- Step 2: Before fabricating composite sheet, measure the surface dimension and the mass of coir fiber sheet.
- Step 3: After fabricating composite sheet, cut the composite sheet according to the surface shape of the original coir fiber sheet. Then measure the mass of cut composite sheet.
- Step 4: The mass of epoxy resin in composite is the difference between the mass of composite sheet in step 3 and the mass of coir fiber sheet in step 2. With the measured density of epoxy resin in step 1, calculate the volume of epoxy resin.
- Step 5: Cut the composite sheet into small specimens and then measure their thickness and deduce the average thickness of the composite sheet, thereby calculating the volume of the composite sheet.
- Step 6: The volume of the composite sheet in step 5 minus the volume of epoxy resin in step 4 is equal to the volume of the coir fiber sheet.
- Step 7: From the mass of coir fiber sheet in step 2 and volume of coir fiber sheet in step 6, calculate the density of coir fiber.

Table 3 presents the measured density of epoxy resin, coir fiber and composite.

Table 3: The density of epoxy resin and coir fiber

Specimens	Epoxy	Composite 1		Composite 2	
		Coir fiber	Composite	Coir fiber	Composite
$\rho\ (g/cm^3)$	1.15	0.98	1.13	0.93	1.12
Reference [6-10]	1.1 – 1.4	1.2		1.2	

From the measured weight fraction and density of component materials with assumption that the volume of void is null, the volume fraction of components in composite groups is calculated and presented in Table 4.

Table 4: Volume fraction of components in composite groups

Specimens	Composite 1		Composite 2	
	Coir fiber	Epoxy	Coir fiber	Epoxy
$V(\%)$	9	91	15	85

5.2. Epoxy resin

For epoxy resin, tensile tests were performed on 3 specimens, which have similar thickness. Fig 6 shows three curves of stress-strain of three epoxy resin specimens obtained from the signals of loadcell and strain gauge. It is observed that three curves are superposed.

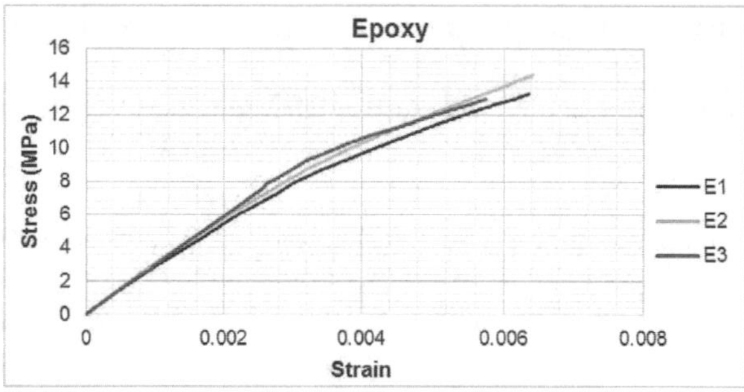

Fig. 6: The curve of stress-strain of epoxy resin specimens

From these curves, in the elastic linear region, the Young's modulus of the epoxy specimens is determined as in Fig. 7.

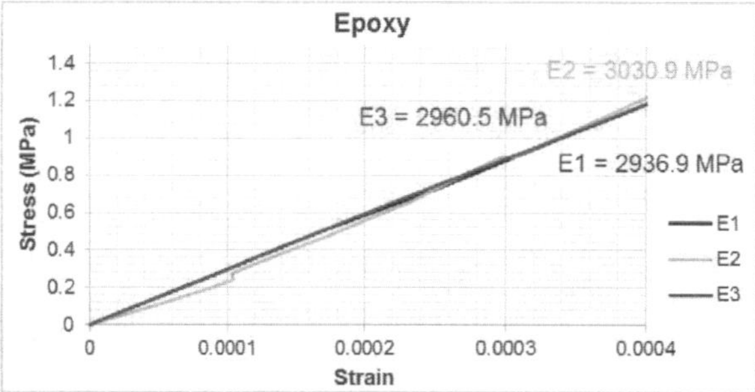

Fig. 7: Linear elastic region of epoxy specimens

For non-linear region, the material has somewhat reduced mechanical properties, so all specimens tend to be highly deformed with a small corresponding force change. Besides, there is a large difference between the specimens in this region and the reasons explained from the effect of flaws in the specimen as well as the effects in the process of setting up experiments such as breaks within the grips, specimen slippage.

For this group, resolution of camera is 2000x3000 pixels which corresponds to 230 pixels for gauge length 10 mm between marks. This resolution is fine enough to measure Poisson's ratio by DIC method. In addition, the theory of isotropic elastic materials is used to determine the shear modulus of epoxy resin. The properties of epoxy resin are given in Table 5.

Table 5: The properties of epoxy resin from experiments and theory

Specimens	**Epoxy resin**	Standard deviation	Reference [8-10]	Error (%)
$E(MPa) = \frac{\sigma}{\varepsilon}$	2976	48.9	2800 – 3400	4.1
$\upsilon = -\frac{\varepsilon_x}{\varepsilon_y}$	0.31	–	0.3	3.3
$G(MPa) = \frac{E}{2(1+\upsilon)}$	1145	–	1192	3.9

In comparison with the values from reference, the difference is small (less than 5%). So, it can be concluded that the experimental results of epoxy resin group are reliable.

In addition, in the non-linear regions, there are other important properties such as yield strength, tensile strength and fracture. However, because of the properties of all materials in this study as well as the limit of maximum applied force of the testing machine, all those properties are not obtained.

5.3. Composite groups

The first group are made from split coir fiber sheet with volume fraction of fiber calculated about 9%. Similar to the epoxy resin group, tensile tests were performed on 3 specimens, which have the fiber distribution higher than the remaining specimens. Fig. 8 shows three curves of stress-strain of three 9% fiber composite specimens obtained from the signals of loadcell and strain gauge. It is observed that three curves are superposed.

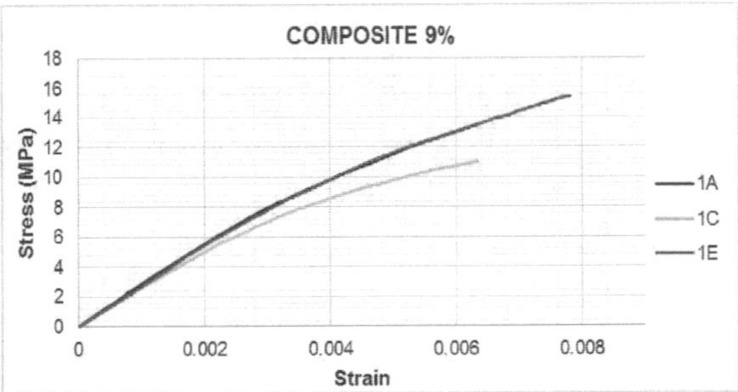

Fig. 8: The curve of stress-strain of 9% fiber composite specimens

For this group, resolution of camera is also 2000x3000 pixels, but gauge length 10 mm between marks just corresponds to 150 pixels. This resolution is not fine enough to measure Poisson's ratio by DIC method.

For the second composite group, the specimens are made from non-split coir fiber sheet with a denser distribution of fibers than the 9% fiber composite group, so it is expected to show the correct properties of the material. Therefore, tensile tests were performed on 6 specimens, which have the volume fraction of fiber about 15%. Fig. 9 shows six curves of stress-strain of six 15% fiber composite specimens obtained from the signals of loadcell and strain gauge. It is observed that six curves are also superposed.

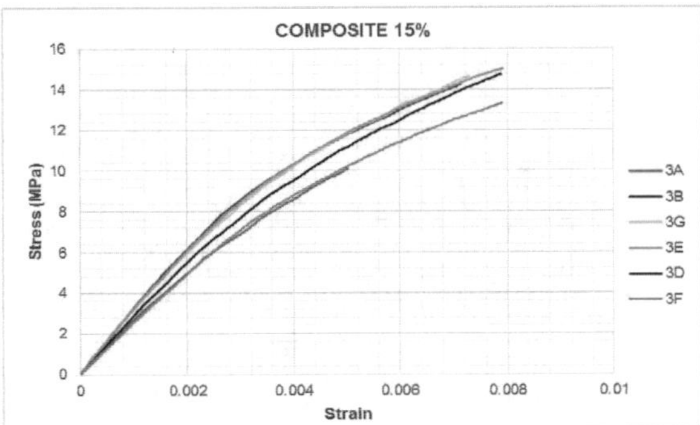

Fig. 9: The curve of stress-strain of 15% fiber composite specimens

For this group, resolution of camera is also 2000x3000 pixels, but in this case, gauge length 10 mm corresponds to 460 pixels. This resolution is really fine to measure Poisson's ratio by DIC method.

The properties of two coir fiber sheet composite groups are given in Table 6.

Table 6: The properties of coir fiber composites from experiments

Specimens	9% fiber composite	15% fiber composite
$E_{11}\ (MPa) = \frac{\sigma}{\varepsilon}$	2726	3168

Standard deviation	135	395
$\upsilon_{12} = -\frac{\varepsilon_x}{\varepsilon_y} = -\frac{\varepsilon_2}{\varepsilon_1}$	–	0.28

Additionally, the rule of mixtures allows to determine mechanical properties of the split and non-split coir fiber sheets as in Table 7. Because of the lack of Poisson's ratio of 9% fiber composite group, for the split coir fiber sheet, just the Young's modulus is defined.

Table 7: The mechanical properties of coir fiber sheet from RoMs theory

	Split coir fiber sheet	**Non-split coir fiber sheet**	**Reference** [4,6-7]
$E_f = \frac{E-(1-f)E_m}{f}$	0.517 GPa	4.251 GPa	4 – 6 GPa
$\upsilon_f = \frac{\upsilon_{12}-(1-f)\upsilon_m}{f}$	–	0.11	
$G_f = \frac{E_f}{2(1+\upsilon_f)}$	–	1915	

The results show that the split coir fiber sheet does not reinforce the mechanical properties of the composite material but reduces them. The cause maybe from factors such as fiber distribution, fiber orientation and composite manufacturing process. Therefore, the mechanical properties of split coir fiber sheet calculated in theory are also very low.

Besides, it can be concluded that the non-split coir fiber sheet shows the similar properties with ones of single coir fiber and results in a higher mechanical performance compared to the manufacture of composite materials from split coir fiber sheet. Additionally, with the result of Young's modulus of coir fiber sheet, it can be concluded that the Young's modulus of single coir fiber will be greater than 4251 MPa.

5.4. Comparison and validation of results

Firstly, to evaluate the reliability of strain measurement, strains measured by two different methods (strain gauge and DIC) in the 15% fiber composite group are compared. The results show that the strain from the DIC (points in Fig 10) is near to the resulting curve from the strain gauge. So, it can be concluded that the strain results from both measurement methods are reliable.

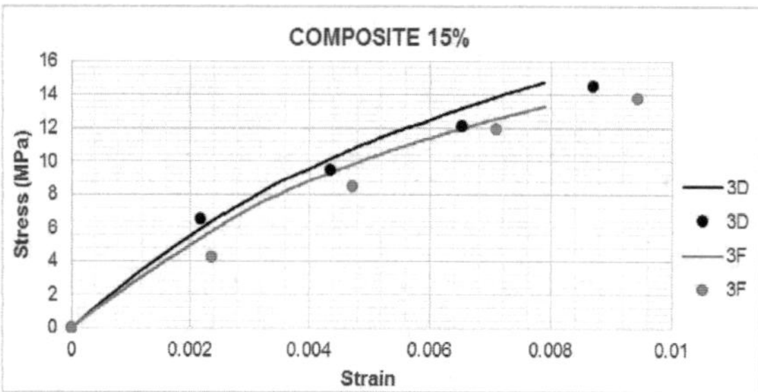

Fig 10: The comparison of two strain measurement methods

Because the composite materials in this study are made of coir fibers in form of sheet and having random orientation, they are not only an orthotropic composite as usually but also considered to be in-plane isotropic elastic material. Therefore, their shear modulus can be calculated in two different ways corresponding for two material behaviors. Table 8 shows the shear modulus of 15% fiber composite calculated from two different ways. The difference between two values of shear modulus is small (less than 5%), it can be concluded that the experimental results of 15% fiber composite group are reliable.

Table 8: Comparison of shear modulus of 15% fiber composite

Specimens	15% fiber composite
$G_{12}(MPa) = \dfrac{E}{2(1+\upsilon_{12})}$	1257
$G_{12}(MPa) = \left(\dfrac{f}{G_f} + \dfrac{1-f}{G_m}\right)^{-1}$	1218
Error (%)	3.1

6. Conclusion

Experimental analysis of the coir fiber sheet reinforced epoxy resin composites is realized in this study. Because of the limit of maximum applied force of test machine, only the Young's modulus, the Poisson's ratio and the shear modulus are determined for composites and component materials. The results show a medium stiffness, small density and low-cost composite material. The improvements like fiber treatment, fiber distribution and orientation as well as manufacturing technology of composite material could help to raise the properties of coir/epoxy composites. Moreover, the coir fibers in Vietnam are really common and affordable. Therefore, the study of mechanical properties of coir fiber composite opens the possibility of application in many fields, even aeronautical technology.

Acknowledgement

The authors would like to thank the Research Fund of Ho Chi Minh City University of Technology for its financial support to the students of PFIEV "French Vietnamese Training Program of Excellent Engineers" in the research project PFIEV_2018-KTGT-02.

References

[1] J. Seena. A comparison of the mechanical properties of phenol formaldehyde composites reinforced with banana fibres and glass fibres. Composites Science and Technology 2002; Volume 62 (14): 1857-1868.
[2] Pickering. A review of recent developments in natural fibre composites and their mechanical performance. Applied Science and Manufacturing 2015; 82: 98-112.
[3] Marimuthu and Chandramohan. A Review on Natural Fibers. IJRRAS 2011; 8: 194-206.
[4] Sathishkumar. Characterization of natural fiber and composites - A review. Journal of Reinforced Plastics and Composites 2013; 32 (19): 1457-1476.
[5] Ku. A review on the tensile properties of natural fiber reinforced polymer composites. Elsevier, 2011; 42 (4): 856-873.
[6] Ali and Majid. Coconut Fibre – A Versatile Material and its Applications in Engineering. Journal of Civil Engineering and Construction Technology 2011; 2(9): 189-197.
[7] Anupama. Experimental Testing of Polymer Reinforced with Coconut Coir Fiber Composites. IJETAE 2014; 4 (12): 453-460.
[8] Campus Plastics. Online 2019. Available: www.campusplastics.com. [Accessed 25 May 2019].
[9] Polymer Database. Online 2019. Available: www.polymerdatabase.com. [Accessed 25 May 2019].
[10] Good Fellow. Online 2019. Available: www.goodfellow.com. [Accessed 25 May 2019].

IOP Conf. Series: Materials Science and Engineering **642** (2019) 012008 doi:10.1088/1757-899X/642/1/012008

Investigation of aircraft engine performance utilizing various alternative fuels

Mahammadsalman Wariman[1], Azami M.H[2], Mark Savill[3], Yi-Guang Li[4], S. A. Khan[5], Ahmad Faris Ismail[6]

[1,2,5,6]Department of Mechanical Engineering,IIUM,Kuala Lumpur,Malaysia
[3,4] Centre for Propulsion Engineering, Cranfield University, United Kingdom

Corresponding author e-mail:hanafiazami@iium.edu.my

Abstract. The airlines are subjected to the energy crisis and have raised environmental issues at the same time. Future engine technology advances could decrease the effect on the environment and energy consumption. Alternative fuelspotentially assist in the reduction of engine emissions and hence lower the energy-related issues. This study presents analysis of the efficiency of aircraft engines as a function of thrust force, flow of the and specific fuel consumption (SFC) at distinct mixing ratios (40% and 100%)of African natural gas, Algae, Camelina, Jatropa, Diesel, Hydrogen, Synthetic paraffinic kerosene, UK natural gas at cruising altitude. In – house Cranfield University simulation codes, PYTHIA & TURBOMATCH have been used to investigate and model a three-shaft high bypass engine analogous to RB211 - 524. The engine model has been certified and authenticated in commercial aircraft with open works found in the Bio - Synthetic Paraffinic Kerosene test program.Blended fuel of Kerosene & hydrogen (KE+HY) fuels gives values of 331.6 KN,1.2577KG/S, and 6.9512 kg/kwh for net thrust force,the flow of fuel and specific fuel feastingat mingling ratio of 40 % respectively. However, at mixing ratio of 100% Blended fuel of Kerosene & hydrogen (KE+HY) fuels gives values of 339.01 KN, 0.800KG/S, and 4.333 kg/kwh for net thrust, fuel flow, and specific fuel consumption respectively.It is found that blended fuel of Kerosene &hydrogen (KE+HY) fuels give better engine performances as compared to other alternative fuels. However, Kerosene &diesel (KE+DI) fuels have shown a slight reduction in engine performance.

1. Introduction

A biofuel is a fuel material derived from renewable biomass resources that are widely used as an alternativ e, cleaner fuel [1].It isdescribed as a fuel composed of long-chain mono-alkyl esters of renewable fatty acids which can be generated by a chemical procedure called transesterification [2].Various liquid biofuels such as biodiesel, bio-ethanol, and bio-oil can be produced from biomass. Biomass was converted into distinct kinds of fuel using thermo-chemical and bio-chemical paths [3].Bio jet fuels consist of a blend of C9–C16 hydrocarbons typically created through transesterification and later hydroprocessing of plant and animal oils to generate fuel with many of the characteristics of petroleum-derived jet fuels [4]. Jatropha, Camelina, and Algae are promising plant-based feedstocks for future aviation biofuels [4].Microalgae have a significant stake in producing biofuels of the next generation that are indistinguishable based on their characteristics from petroleum fuels. This is because the use of microalgae to produce biofuels would have less adverse effects on food supply and other agriculture as they can be grown on non-arable soil using fresh, waste or saline water sources and allow more effective recycling of nutrients and greater productivity [7]. Green (Chlorophyta), red (Rhodophyta), and brown algae (Phaeophyta) are the most prevalent form of algae. Because of the high lipid content, green microalgae are more appropriate for biofuel manufacturing than other microalgae [8].As we understand, aircraft use has proliferated over the past few years. More than six billion individuals travel around the globe each year by aircraft [5].In the past decade, there has been a huge growth of technology in the direction of alternative fuels. Aircraft can be fueled with synthetic jet fuels or jet biofuels nowadays [6].However, the demand for fuel in the world has risen gradually every day, owing to fast population growth [7]. According to literature, petroleum consumption in 2012 was estimated at 89 million barrels per day across the globe, and nearly half of this was used in gasoline manufacturing [8]. The oil resources are expected to run out at this pace in the next 50 years [8].H. D. Banu *et al.*[9].Reported that at current rates of consumption, it is estimated at a 95% probability that the worlds remaining oil resources will last 63 years and a 5% chance that they will last 95 years. Research on alternative aircraft fuels is therefore of paramount significance for the long term [9].As oil fossil fuel decrease or a significant carbon tax on fossil fuels is imposed globally, liquid fuels from biomass-based on carbon will become widely competitive [10]. Growth in fuel usage rises by around 2.9% annually, in aviation fuels by around 4%[11]. It is projected that crude oil production will be maximum in 2020. Total emissions of carbon dioxide are expected to rise by more than 60%[11].Depleting fossil fuels and pollution induced by the combustion of petroleum base makes the world look to renewable energy as an alternative fuel to meet energy demand for economic development [12].The aviation sector is researching fuel availability issues in response to increasing environmental needs [13]. It is advisable to generate energy from renewable sources such as solar, hydro, tidal, wind, biomass and geothermal power that do not produce environmentally pollution and are accessible in large quantities [14].One of the most significant study subjects is the search for an alternative fuel independent of carbon-based fossil fuels [15]. Alternative fuels were suggested as one of the alternatives to decrease the greenhouse gas (GHG) footprint of aviation [16].It is possible to convert agricultural, industrial and domestic waste into biodiesel, biogas using various methods [17].Biofuel's usefulness in aerospace fuel these days owing to biofuel characteristics. Biofuel will be commonly used in the near future as an aviation fuel [18].Biofuels are other alternative kinds of fuel to decrease worldwide emissions [5]. These energy resources give a number of benefits over non-renewable fossil fuels [8].The benefits of using biofuels would be their balanced effect CO_2on the environment and their ability to become a sustainable fuel.Their use can also lead to reduced emissions from the engine.If sustainability, efficiency, and cost liabilities can be overcome, the aim is to mix biofuels with synthetic jet or Jet-Aviation fuels [19].Synthetic jet fuels are produced from coal, gas, or other hydrocarbon feed stocks using a Fischer-Tropsch process. The efficiency of these fuels is very comparable to standard jet fuel, but they contain almost zero sulfur and aromatics. This may lead in reduced exhaust emissions. Furthermore, synthetic fuels have outstanding low temperature characteristics, keeping low viscosity at reduced ambient temperatures[19]. The primary motivation for this research has provided rise to a desire for more secure fuel supplies for aircraft in the future from renewable and sustainable energy sources obtained from renewable fuel sources.

2. Previous works on alternative fuels

Previous laboratory and ground test experiments using bio-based fuels or synthetic Fischer – Tropsch fuels generated from natural gas and coal feedstocks indicate that the lack of sulfur and aromatic species in the fuel significantly decreases airplane engine sulfate and black carbon emissions[4].M. Z. W. Yahya *et al.*analyzed liquid hydrogen and found that hydrogen is better than kerosene fuel. The hydrogen performs better in term of performance and also in reducing overall emission [20].M. H. Azami *et al.* proved that biofuels based on jatropha are one of the most potential plants in biofuels [21].A. J. Beyersdorf *et al.* used conventional oil JP-8 fuel, pure synthetic fuels generated from natural gas and coal feed stocks using the Fischer–Tropsch (FT) method, and 50% of both fuel mixtures were tested on a DC-8 airplane in the CFM-56 motors. A positive effect of alternative jet fuels is a reduction in particulate and gas emissions [22].M. H. Azami *et al.* showed that Jatropha biofuel works much better in gross thrust, fuel flow, and SFC than camelina biofuel[22].X. Hui *et al.* in there research utilized conventional Jet-A and six alternative jet fuels, including three Fischer – Tropsch Synthetic Paraffinic Kerosene (SPK) fuels and three Hydrotreated Renewable Jet (HRJ) fuelsand tested to acquire their basic combustion properties[23].R. O. Price *et al.* researched the use of liquid hydrogen as a replacement commercial aviation fuel offering enhancement in air quality [24].Two kinds of fuel were tested by B. Gawron *et al.* Aviation fuel (Jet A-1) and a combination of aviation fuel with HEFA synthetic element (hydro-processed esters and fatty acids) from camelina feedstock [25].S. B. Amgad *et al.* showed that FT jet fuel from natural gas, coal, and biomass.Bio-jet fuels from fast pyrolysis ofbiomassand hydro processed renewable jet fuel from vegetable and algal oil reduced emission[26].Mohdnoh *et al.* reviewed that enhanced composition in bio-jet fuel to be compared with conventional fuel creating more advantages in term of carbon emissions, fuel efficiencies, components wear andtear and reduction in fuel cost and maintenance in prolong future [27]. Fig.1displays the relative CO_2 emissions generated by different fuels during their life cycles, using present jet fuel as the basis. FT fuels can only be regarded a feasible alternative to petroleum if the CO_2 emissions produced during manufacturing can be captured and sequestered continuously [19].Achieving deep CO_2 reductions will also require a shift to very low-carbon energy carriers, of which the three most likely to play a prominent role are electricity, biofuels, and hydrogen [28]. Sustainable bio-jet fuels are a promising route for mitigating greenhouse gas emissions [4].

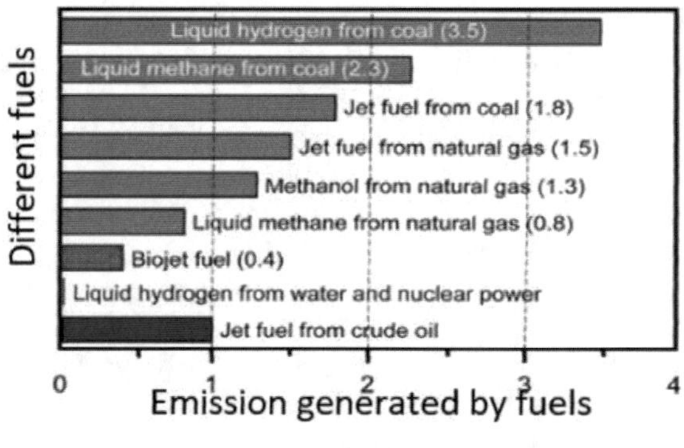

Fig. 1: Co_2 emission by different fuels (JET A) [19].

3. Present work

This research emphasizes the effect of mixed alternative fuels on the effectiveness of aircraft engines specifically on gross thrust, and specific fuel ingestion (SFC) at distinctly mixed mixing ratio percentages. Alternative fuels such as African natural gas, Algae, Camelina, Jatropa, Diesel, Hydrogen, Synthetic paraffinic kerosene, UK natural gas biofuel are evaluated blended with kerosene (C_12 H_24) at 40% and 100%. Engine model RB211-524 was used for verification during the evaluation using available engine constraints.Authentication of a variant RB211 and comparison is done with Rahmes et al. research work [29]. For computational analysis, our native code was used.PYTHIA uses a modified Newton-Raphson convergence method is used for designing and PYTHIA is also used in the zero-dimensional steady-state modelto compute numerous gas turbine devices for both design and off-design themes.It can also assist as ananalyticaldevice for examining deterioration and enables map scrambling for off-design situations. PYTHIA is united with our actassessment program TURBOMATCH. PYTHIA demands for the TURBOMATCH program FORTRAN-coded to reiterate the energy & mass equilibriumof every engine constituent. PYTHIA is easy to use and has a newengine component choicefor the interface. PYTHIA is the abilityofindustrialized gas turbines to aero-gas turbines has been assessed &validated for many years. The newestform of PYTHIA can change the type of fuel &the ratio of amalgamatedfraternization ratio while preserving the same engine design as the conventional kerosene.This is crucial if fit for use fuels for the actual engine at different working points are to be evaluated. This research can, therefore, serve as an expansion of work of Azami, M.H.*et al.* [30]. Work using a previous form of PYTHIA that for original design conditions could only provide comparison for separate pure fuels. These latest findings not only support earlier outcomes but also go beyond them as more abilities have been put in new PYTHIA variant, such as assessing added alternate fuel choices under separate off-design circumstances.

4. Methods

The RB211-524 engine is selected for research. From library data of PYTHIA selected engine configuration was specified and arrangements made for defaulting situations. The classical engine arrangements are shown in Fig. 2. Kerosene is selected as a baseline fuel. Each engine model component is labeled as a brick and has its own functionality. The inputs were allocated in the INTAKE brick, agreeing to the suggested flying requirements such as altitude, flight velocity, mass flow, pressure recovery, deviation of pressure, and relative humidity. For the pressure ratio is 2.0, and the stator angle is 10o for the first compressor, the maximum for the first compressor, the maximum pressure ratio is 2.0, and the stator angle is -10 °. The previous HP compressors have a maximum pressure ratio of 11.0 with stator angle of -10o. But it is anticipated that only the HP compressors will have bleeding air. By setting the complete relate variations of mass movement and overall pressure, PREMAS bricks are used to determine the outlet circumstances of parts such as splitters, bleeds, bypass ducts or jet pipes. The burner does not have water flow. In the meantime, MIXEES brick is used to determine the outlet circumstances causing from steady two-flow blending without the total pressure loss allowance & after TURBINE brick information, MIXFUL brick data is used to calculate outlet requirements from steady mixing area flows with total allowance for complete pressure shift due to momentum equilibrium. The 0.04 ratio of the maximum enthalpy drop and the turbine inlet temperature for all turbines is set at 1580K. These turbines also have a 10o angle position and are choked at low velocity. The NOZZLE brick selects a convergent nozzle. Tabulated in Table 1, the engine parameter results, and the baseline fuel efficiency. The PYTHIA method flowchart is presented in Fig. 3. It starts with inputs defined by the user. TURBOMATCH is called for reiteration in mass & energy balance relation (i.e., Eq. 1 & 2). Equations must and should be contented between consecutive apparatuses. Before the iteration method, values for pressure ratio, temperature, and rotation swiftness must be guessed. A high-level computer programming language FORTRAN is used to code TURBOMATCH. TURBOMATCH is Cranfield's software for Gas Turbine Performance Simulation. For the method of mass balance iteration, compressor and turbine maps were required. For the assessment of thermochemical fuel characteristics, NASA Chemical Equilibrium Analysis (CEA) is used. In the TURBOMATCH library information, these correlations are stored. Before the iteration method converges, several guess values will be required. Finally, the data were performed and submitted for data analysis to the Excel spreadsheet. Table 1 shown the design condition for the engine parameters and performance utilizing kerosene fuel as the baseline. These values are obtained through iterations for every brick used in the schematic diagram (Fig. 2).

IOP Conf. Series: Materials Science and Engineering **642** (2019) 012008 doi:10.1088/1757-899X/642/1/012008

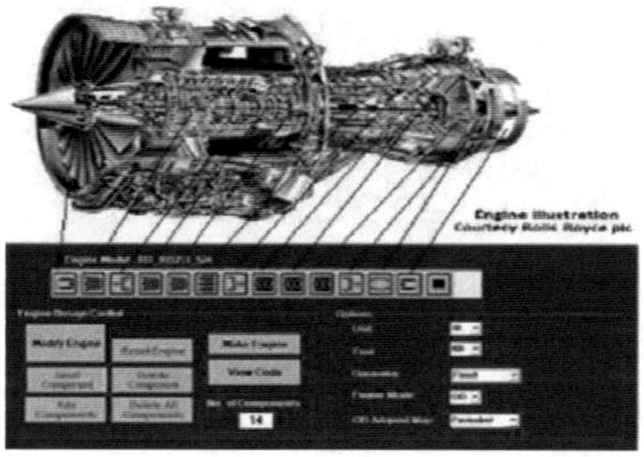

Fig. 2: PYTHIA engine model schematic diagram.

Table1. Engine performance and parameters for baseline fuel.

INTAKE		COMBUSTORS		TURBINE				NOZZLE		ENGINE PERFORMANCE	
Altitude (m)	10588	Efficiency	0.99		1	2	3	Area (m²)	2.25	Bypass Pressure Ratio	4.3
Intake of mass flow (kg/s)	670	Drop in pressure (atm)	1.29	Efficiency	0.91	0.92	0.92	Exit Velocity (m/s)	394.0	Gross Thrust (kN)	293.38
Relative Humidity (%)	60	Flow of fuel (kg/s)	2.18	T_{total}(K)	1580	1499	1240	Nozzle Coefficient	0.98	Specific Thrust (N/kg. s)	154.71
Momentum Drag (kN)	189.72	LHV (MJ/ kg)	43.12	P_{total}(atm)	31.04	31.04	12.44	T_{total}(K)	464.39	Fuel Flow (kg/s)	2.18
Flight Mach Number	0.84	P_{total}(atm)	31.04	Mass flow rate (kg/s)	112.18	128.61	128.61	P_{total}(atm)	1.58	SFC (kg/N. s)	21.07
		Fuel-to-air Ratio	0.02								

$$\frac{W_n\sqrt{T_n}}{P_n} = \frac{W_{n+1}\sqrt{T_{n+1}}}{P_{n+1}} \quad (2)$$

(1) $$TurbineWork(TW) = CompressorWork\ (CW)$$

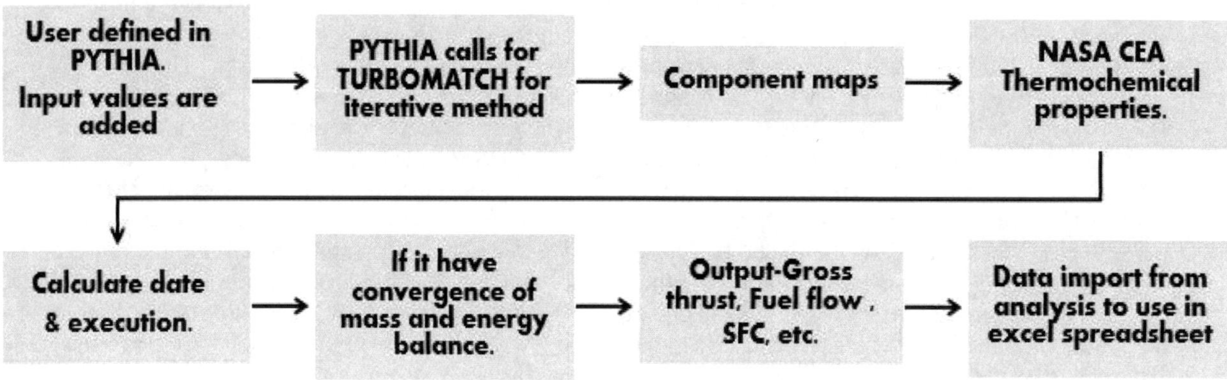

Fig. 3.PYTHIA data process flowchart.

IOP Conf. Series: Materials Science and Engineering **642** (2019) 012008 doi:10.1088/1757-899X/642/1/012008

5. Results

Fig.4 shows a list of legends used for all graphs described below.

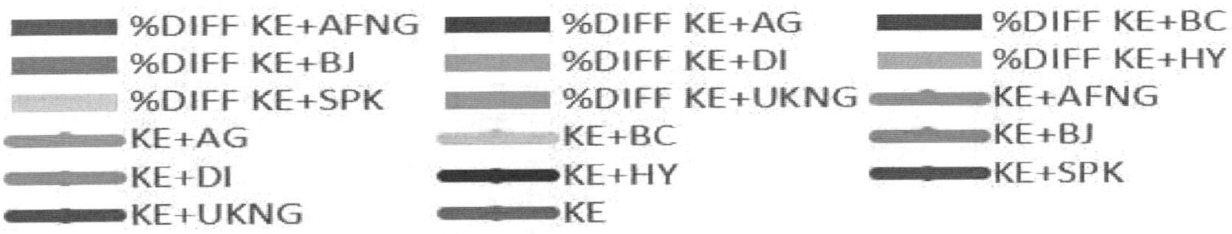

Fig. 4: Legends considered for all graphs shown in paper

5.1. Varying flight conditions

The INTAKE block diagram is modifiedto define the flight conditions difference in off-design automatically.This is achieved by changing the speed of the flight. PYTHIA is programmed for various flight circumstances at off-design points.

5.2. Gross thrust at various mixing ratio at different Mach number

This portion describes the impacts of mixing 40% and 100% on engine enactment at the sailingcondition. As illustrated in below figures which shows the difference in gross thrust and proportionvariance as related to the baseline fuel. As shown in Fig5. KE+HY shows the maximum value of 331642.4 N gross thrustsmonitored by KE+UKNG and KE+BJ320760.2 N and 317844.7 N at Mach number 0.8 The minimum value of 212735.5N gross thrust shown by KE +DI. And, there is no much gross thrust value difference between KE+AG, KE+BC, KE+BJ. At this gross thrust, there is no considerable percentage difference of fuel with respect to KE.In Fig. 6, the KE+HY shows the maximum value of 339018.66 N gross thrusts followed by KE+UKNG and KE+AFNG of value 325625.4 N & 322372.99 N. The minimum value of gross thrust is shown by KE +DI of value 212554.1N. And, there is no much gross thrust value difference between KE+AG, KE+BC, KE+BJ. In all mixing ratios percentage difference with respect to KE goes on increasing as speed increases. Finally, KE +HY shows the highest increment of gross thrust at a speed of 0.8.

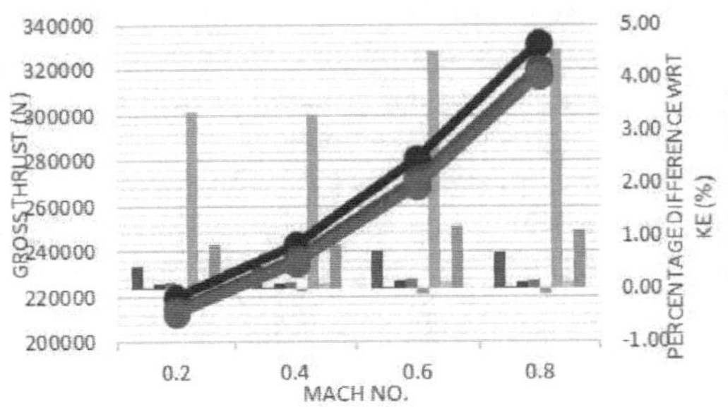

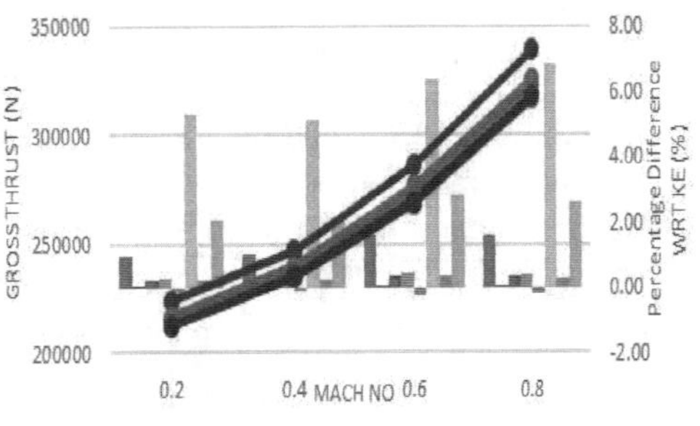

Fig.5: Variation of gross thrust at 40% mixing ratios.

Fig.6: Variation of gross thrust at 100% mixing ratios.

5.3. Fuel flow at various mixing ratio at different Mach number

IOP Conf. Series: Materials Science and Engineering **642** (2019) 012008 doi:10.1088/1757-899X/642/1/012008

This portion describes the impacts of mixing 40% and 100% on engine enactment at conditions the sailing flight. The ambient and flight circumstances are shown in the below figures, which shows fuel flow difference and fraction variance as related to the standard fuel. As shown in Fig.7. The maximum value of 2.2756KG/S fuel flows at the mixing of 40% shown by the KE+DI followed by KE+AG 2.2696KG/S. The lowest value of fuel flow is by KE +HY and KE+AFNG of value 1.2577& 1.933KG/S.

Fig.8 shows the Maximum value of 2.2888KG/S fuels flows at the mixing of 100% as shown by KE+DI followed by KE+AG 2.2737KG/S. The lowest value of fuel flow is by KE +HY 0.8009 KG/S and KE+AFNG 1.8031KG/S of value. In both different mixing ratios, fuel flow is almost linear at various speeds. There is a slight decrease in fuel flow at the highest speed flow of 0.8. And, the percentage difference with respect to KE goes on increasing as the mixing ratio increases.

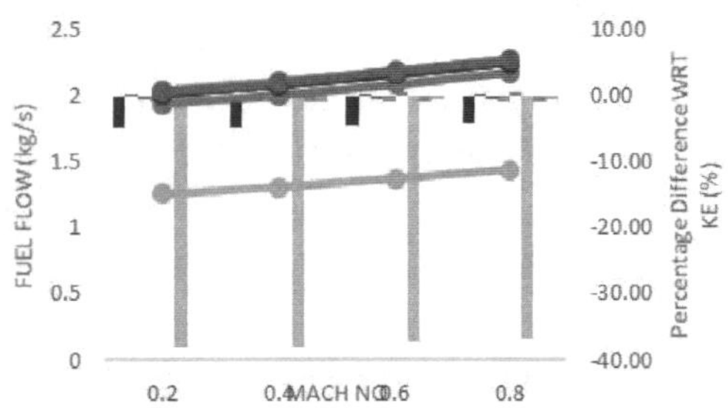

Fig. 7: Variation of fuel flow at 40% mixing ratios.

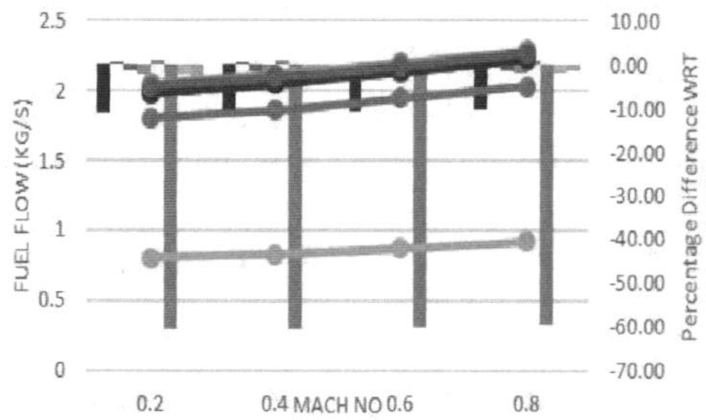

Fig. 8: Variation of fuel flow at 100% mixing ratios.

5.4. SFC atvarious mixing ratio at differentMach number

This portion describes the impacts of mixing of 40% and 100% on engine show at conditionsof the sailing flight. The ambient and flight situations are shown in the below figures, which present the specific fuel consumption deviation and proportion difference as related to the standardfuel. As shown in Fig. 9, the maximum value of SFC Shown by KE +DI 18.4338 KNfollowed by KE+AG18.3593KN&KE+BC 18.1914 KN. The lowest value of 6.9512 KN SFC shown by KE+HY.And, as shown in Fig. 10. The maximum value ofSFC Shown by KE +DI18.5775 KN followed by KE+AG18.39 KN &KE+BC17.9758KN of value. The lowest value of 4.3333 KN SFC showed by KE+HY. In both cases, SFC goes on increases as speed increases. There is no much difference in SFC of KE+AFNG, KE+AG, KE+BC, KE+BJ, KE+SPK, KE+UKNG.Test conducted on various biofuel mixtures like African natural gas, algae, camelina, Jatropa, diesel, hydrogen, synthetic paraffinic kerosene, UK natural gas at an altitude of 2000 feet. Table 2. shows in brief results of fuels which given highest &lowest values for gross thrust, fuel flow, specific fuel consumption. From this work, it is investigated that blended fuel of Kerosene & hydrogen is the best alternative fuel option and Kerosene + diesel not shown good results for net thrust, and specific fuel ingesting. To generate more gross thrust 100% mixing of Kerosene & hydrogen is recommended. For fuel flow & specific fuel feeding, 40% mixing of kerosene & hydrogen gives excellent results comparatively.

Table 2. Maximum &minimum values of gross thrust, fuel flow &specific fuel consumption with respect to base line fuel.

Performance Parameters	Blended Ratio Percentage	Maximum diff in % from base line fuel		Minimum diff in % from base line fuel	
Gross thrust (KN)	40%	KE+HY	4.51	KE+DI	-0.09
	100%	KE+HY	6.83	KE+DI	-0.22
Fuel flow KG/S	40%	KE+HY	-38.04	KE+DI	0.42
	100%	KE+HY	-60.54	KE+DI	1.06
Specific fuel consumption (SFC) kg/kwh	40%	KE+HY	-40.51	KE+DI	-0.52
	100%	KE+HY	-63.62	KE+DI	1.3

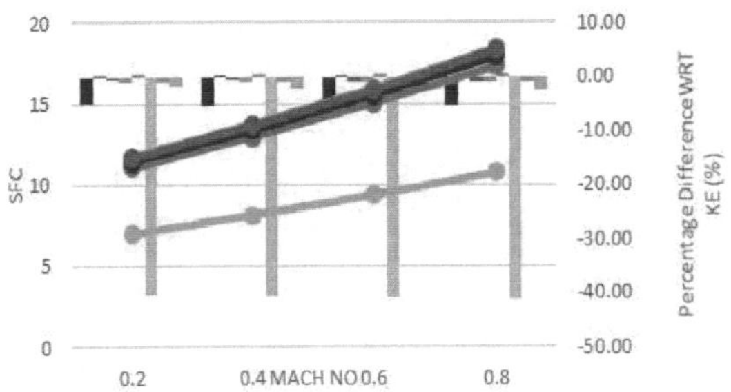

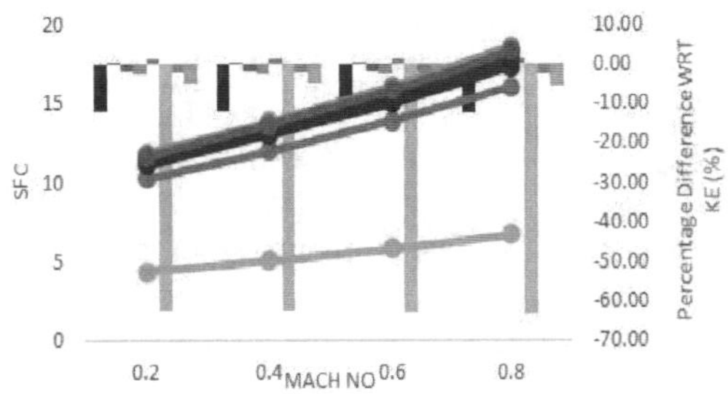

Fig. 9: Variation of SFC at 40% mixing ratios. **Fig. 10:** Variation of SFC at 100% mixingratios.

6. Conclusion

The present study has investigated the impact of mixed biofuels on the efficiency of airplane engines in particular on net thrust, the flow of fuel and specific fuel consumption (SFC) at a different speed and at two fraternization ratios. Three biofuels-Algae, Jatropha and Camelina are assessed as a clean fuel and mixed with kerosene($C_{12}H_{24}$) at 40% and 100%. The outcomesshowed that the reduced fuel heating value had a significant impact on engine such as thrust, fuel flow, and SFC at variousblended mixing ratio at altitude 2000 feet. Pure alternative fuels for two mixingratios at altitude 2000 feet were evaluated. KE+HY fuels showed a considerablyimproved engine performance as correlated to all the biofuels. KE+DIfuels presented a very less engine enactment asmatched to all the biofuels.

Acknowledgment

The corresponding author would like to thank and acknowledge International Islamic University Malaysia for the FRGS Grant (FRGS19-063-0671) for this research project.

References

[1] Costa JF, Almeida MF, Alvim-ferraz MCM, Dias JM. Biodiesel production using oil from fish canning industry wastes. Energy Convers Manag 2013;74:17-23. doi:10.1016/j.enconman.2013.04.032.

[2] Giakoumis EG, Rakopoulos CD, Dimaratos AM, Rakopoulos DC. Exhaust emissions of diesel engines operating under transient conditions with biodiesel fuel blends. Prog Energy Combust Sci 2012;38(5):691-715. doi:10.1016/j.pecs.2012.05.002.

[3] Fadhil AB, Ahmed AI, Salih HA. Production of liquid fuels and activated carbons from fish waste.

Fuel 2017;187:435-445. doi:10.1016/j.fuel.2016.09.064.

[4] Moore RH, Thornhill KL, Weinzierl B, et al. Biofuel blending reduces particle emissions from aircraft engines at cruise conditions. Nature 2017;543(7645):411-415. doi:10.1038/nature21420.

[5] Baharozu E, Soykan G, Ozerdem MB. Future aircraft concept in terms of energy efficiency and environmental factors. Energy 2017;140:1-17. doi:10.1016/j.energy.2017.09.007.

[6] Zhukov VP, Sechenov VA, Starikovskiy AY. Autoignition of kerosene (Jet-A)/air mixtures behind reflected shock waves. Fuel 2014;126:169-176. doi:10.1016/j.fuel.2014.02.036.

[7] Yesilyurt MK. The effects of the fuel injection pressure on the performance and emission characteristics of adiesel engine fuelled with waste cooking oil biodiesel-diesel blends. Renew Energy 2019;132:649-666. doi:10.1016/j.renene.2018.08.024.

[8] Banu HD, Shallangwa TB, Joseph I, Magu TO, Hitler L, Ahmed S. Biodiesel Production from Neem Seed (Azadirachta indica) Oil Using Calcium Oxide as Heterogeneous Catalyst. J Phys Chem Biophys 2018;8(2):8-10. doi:10.4172/2161-0398.1000266.

[9] Ebbinghaus A, Wiesen P. Aircraft fuels and their effect upon engine emissions. Air Sp Eur 2002;3:101-103. doi:10.1016/s1290-0958(01)90026-7

[10] Openshaw KA review of. A review of Jatropha curcas: an oil plant of unfulfilled promise. Biomass Bioenergy 19 2000;19:1-15.

[11] Pągowski Z. New perspectives for biofuels in aviation. J KONES Intern Combust Engines 2003;10:1-4. http://ilot.edu.pl/KONES/2003/3-4/28.pdf.

[12] Mechanics A. feasibility study of pulse detonation engine fueled by biogas. Appl Mech Mater 2013;388:257-261. doi:10.4028/www.scientific.net/AMM.388.257

[13] Daggett D, Hadaller O, Hendricks R, Walther R. Alternative Fuels and Their Potential Impact on Aviation. In: The 25th Congress of the International Council of the Aeronautical Sciences (ICAS) 2006:1-8. doi:10.1016/j.vetpar.2012.01.033.

[14] Kirubakaran M, Arul Mozhi Selvan V. A comprehensive review of low cost biodiesel production from waste chicken fat. Renew Sustain Energy Rev 2018;82:390-401.

[15] Jun Li, Hongyu Huang, Noriyuki Kobayashi ZH and YN. Study on using hydrogen and ammonia as fuels: Combustion characteristics and NOx formation. Int J energy Res 2013;32(32):336-355. doi:10.1002/er.

[16] Pfaender H, Payan AP, Mavris DN. Long-term Alternative Fuels Scenarios. In: AIAA/3AF Aircraft Noise and Emissions Reduction Symposium AIAA2014:1-18. doi:10.2514/6.2014-3165.

[17] Warimani M, Azami MH, Khan SA, Ismail AF. Study Of Feasibility Of Pulse Detonation Engine Powered By Alternative Fuels. Int J Eng Adv Technol 2019;8(January):291-296.

[18] Shimada T, Hayashi AK, Yamada E, Tsuboi N. Numerical Study on Detonation Characteristics Using a Bio-Fuel. In: 49th AIAA Aerospace Sciences Meeting Including the New Horizons Forum and Aerospace Exposition 2011:1-11. doi:10.2514/6.2011-582.

[19] Daggett Hendricks, R.C., Walther, R., Corporan, E. DL. Alternate fuels for use in commercial aircraft. In: 18th ISABE Conference September 2007:1-21.

[20] Yahya MZW, Azami MH, Savill M, Li Y, Khan SA, Salman M. Modelling of a Three - Shaft High - Bypass - Ratio Engine Performance and Emission Prediction Using Hydrogen Fuels. Int J Innov Technol Explor Eng. 2019;8(6S3):484-490.

[21] Azami MH, Noorazman Z, Savill M, Li YG, Hilmi MR. Modelling the performance and emission prediction of RB211 aero-gas turbine engine fuelled by Jatropha-based biofuel. In: IOP Conference Series: Materials Science and Engineering 2019; 488:1-6. doi:10.1088/1757-899X/488/1/012006.

[22] Beyersdorf AJ, Timko MT, Ziemba LD, et al. Reductions in aircraft particulate emissions due to the use of Fischer – Tropsch fuels. Atmos Chem Phys 2014:11-23. doi:10.5194/acp-14-11-2014.

[23] Hui X, Kumar K, Sung C, Edwards T, Gardner D. Experimental studies on the combustion characteristics of alternative jet fuels. Fuel 2012;98:176-182. doi:10.1016/j.fuel.2012.03.040.

[24] Price RO. Liquid hydrogen-an alternative aviation fuel? Int J Hydrogen Energy. 1991;16(8):557-562. doi:10.1016/0360-3199(91)90155-C

[25] Gawron B, Bia T, Janicka A, Zawi M. Exhaust toxicity evaluation in a gas turbine engine fueled by aviation fuel containing synthesized hydrocarbons. Aircr Eng Aerosp Technol. 2019. doi:10.1108/AEAT-11-2018-0277.

[26] Amgad Elgowainy, Jeongwoo Han, Michael Wang, Nicholas Carter, Russell Stratton, James Hileman, Andrew Malwitz and SB. Life-Cycle Analysis Of Alternative Aviation Fuels In Greet 2012:1-76.

[27] H Mohd Noh GAR and NAAR. The adaptation of sustainable biojet fuels and its effect on aircraft engine maintenance. In: IOP Conference Series: Materials Science and Engineering 2016:1-10. doi:10.1088/1757-899X/152/1/012043

[28] Lewis M. Fulton, Lee R. Lynd, Alexander Körner, Nathanael Greene LRT. The need for biofuels as part of a low carbon energy future. Biofuels, Bioprod Biorefining 2015:1-8. doi:10.1002/bbb

[29] Rahmes TF, Kinder JD, Henry TM, et al. Sustainable Bio-Derived Synthetic Paraffinic Kerosene (Bio- SPK) Jet Fuel Flights and Engine Tests Program Results. In: 9th AIAA Aviation Technology, Integration, and Operations Conference (ATIO)
and
Air 21 - 23 September 2009, Hilton Head, South Carolina2009:1-19.

[30] Azami MH, Savill M. Comparative study of alternative biofuels on aircraft engine performance. Proc Inst Mech Eng Part G J Aerosp Eng 2017;231(8):1509-1521.

Design and Manufacture of Miniature Testing Machine for Composite Materials

Le Nhan NGUYEN*, **Van Quy HO, Song Thanh Thao NGUYEN, Thi Hong Hieu LE**

Department of Aerospace Engineering, Ho Chi Minh City University of Technology, Vietnam

*Corresponding author e-mail: 1412624@hcmut.edu.vn

Abstract. Today, there are many commercial test machines in industry and laboratory with many different sizes. However, their prices are high. Some laboratories made their own equipment at lower price based on their situation, demand and purpose of test. In this paper, the design of the frame structure and the operating systems, as well as the calibration of a small tensile and bending test machine for low- to medium-stiffness composite materials are presented. The construction is made of aluminum. The loading is created by an actuator which is controlled by an Arduino micro-controller. A Hocdelam USB-9090 (the same as that of NI USB-6009 but with less accuracy) data acquisition card is connected to a personal computer through an interface with load cell and strain gauge by the amplifier to record the force and the displacement during a test. The rate of loading is identified by the rate of actuator which is maintained stable thanks to the adjustment of PWM (Pulse Width Module) with Arduino. The machine is able to conduct tensile and bending tests with a loading up to 3000 N with a sensitivity of 1 N. It is possible to conduct tests on composite specimens at speeds of 1.7 ~ 300 mm/min. This machine is able to obtain results for mechanical properties which are in good agreement with the data obtained from a commercial machine, the differences are less than 5%.

Keywords: miniature testing machine, frame structure, specific speed, data acquisition, composite material.

1. Introduction

Nowadays, the material industry is increasing rapidly so that demands of measurement of the mechanical properties of materials are significantly growing up. Hence, various universal test machines are developed for industrial or study purposes in the world. Some laboratories made their test machines based on their purposes of study and their economic conditions. Many researchers are interested in designing miniature testing machines to obtain the mechanical properties of materials [1-4]. Partherpan and his colleagues [1] proposed a simple miniature disc-type tensile specimen and fixtures to hold specimens with the help of a rigid pin to predict the mechanical properties of materials.

In our study, a small tensile and bending test machine for low- to medium-stiffness (Young's modulus is less than 10 GPa) composite materials are presented. The construction of the miniature testing machine is designed as a universal test machine. However, it is on a small scale with a shaped aluminum structure for strong load-bearing ability. The system composing a linear actuator and connectors allows to create a load up to 3000N and to maintain the loading rate stable during the test. The frame also ensures the co-axial movement of the tensile specimen and grips and the linear motion of the actuator.

In terms of data acquisition, a GUI LabVIEW program is written to help users assess the mechanical properties of material generally. Force is recorded by a load cell, strain is measured by a strain gauge and displacement of grips is identified by a linear potentiometer. Because the voltage outputs of the load cell and the strain gauge are very small (few mV), they need to be amplified before they are transferred to a data reading card. As a result, the accuracy of acquired data depends on the resolution of the data collecting card. In our case, a USB-9090 Hocdelam data acquisition card is used. The resolution of this 10-bit Analog to Digital Converter (10-bit ADC) is high enough to obtain accurately the signals of sensors.

2. Design and manufacture

2.1. Operating principle

For a miniature testing machine conducting both tensile and bending tests, the basic components must be the frame, the force generation system, grips for the tensile test and fixtures for the bending test. There are also necessary components such as force, displacement and strain measurement devices. The machine must meet some technical requirements such as loading rate, the coaxial movement and the load-bearing capacity. The operating principle of the miniature testing machine is shown in Fig. 1.

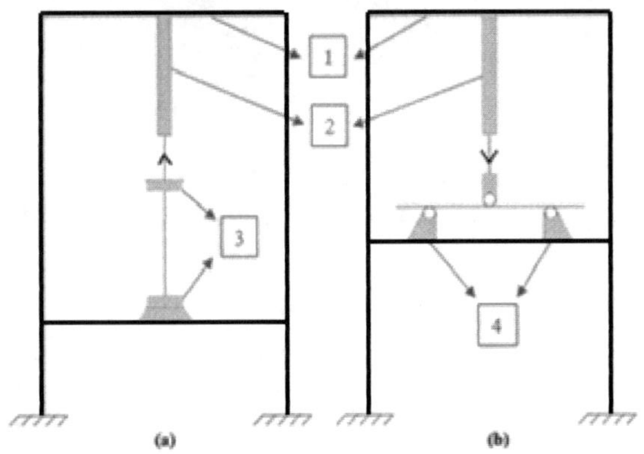

Fig. 1: Operating principle of testing machine in the tensile test (a) and bending test (b) with: 1. Frame; 2. Force generation system; 3. Grips; 4. Fixtures

2.2. Construction

After creating the final design idea of the miniature testing machine, the components are chosen by considering their advantages and disadvantages.

2.2.1. The frame

In terms of construction, the design requires the machinability, the maintainability and the load-bearing capacity. For the frame, the shaped aluminum profiles are selected (Fig. 2). These treated aluminum bars have appropriate mechanical properties and are suitable for assembly structure.

Fig. 2: Shaped aluminum and design idea for the frame

Besides the frame, the design of intermediate parts for installing the force generation component and for ensuring the co-axiality is shown in Fig. 3.

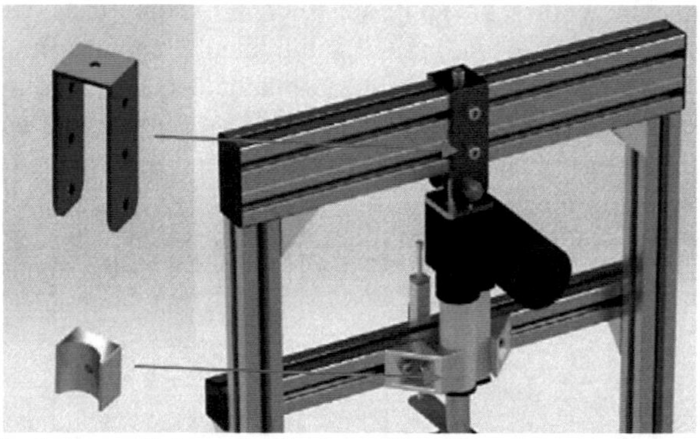

Fig. 3: The intermediate parts for installing force generation component

2.2.2. The force generation system

With outstanding advantages, the linear actuator is chosen as the solution for the force generation component of the testing machine (Fig. 4). This device uses an axial drive mechanism and ensures the standardization for the purpose of easy maintenance, repair and replacement.

Besides, for installing the force measurement device, the designed intermediate connection parts have grooves to ensure that the device does not rotate during a test (Fig. 5). In addition, the grips for tensile test and the fixtures for bending test are attached below the force measuring device as shown in Fig. 6.

All intermediate parts, grips and fixtures are made of C45 steel due to its strength and machinability.

2.2.3. The measurement devices

Firstly, due to the advantages of high accuracy and high standardization, the loadcell (Fig. 7 (a)) is selected as a force measurement solution in the testing machine. Loadcell is a device used to convert force or weight into electrical signals and used to measure large, static or slow-changing forces. Loadcell operates on the principle of Wheatstone bridge circuit. To satisfy technical requirements and limited funding, the selected loadcell can measure a maximum load of 1000 N for both tensile and compression forces.

Next, the strain gauge (Fig. 7(b)) is selected as a strain measurement device in this testing machine. The strain gauge is actually a resistor, whose resistance changes when the length of the resistor changes and thus gives the strain value. Similar to loadcell, strain gauge uses the Wheatstone bridge circuit to ensure high accuracy.

Finally, a linear potentiometer (Fig. 7(c)) is used as a device to measure the displacement of grips in tensile test or the deflection of specimens in bending test. It is similar to a digital absolute indicator but it has outstanding advantages such as higher accuracy and ability to connect with readers to capture data continuously in real time.

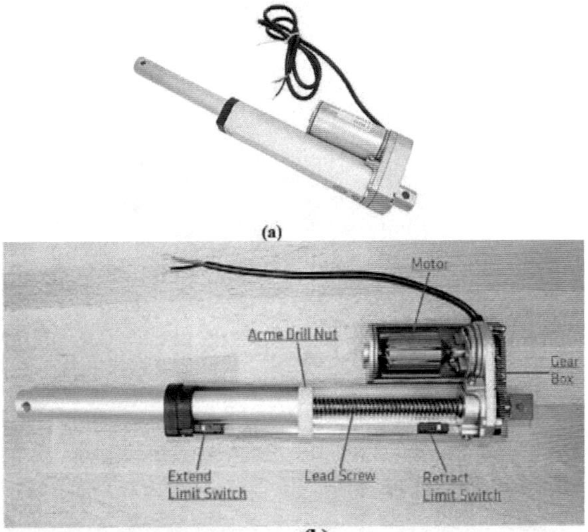

Fig. 4: Linear actuator (a) and its internal structure (b)

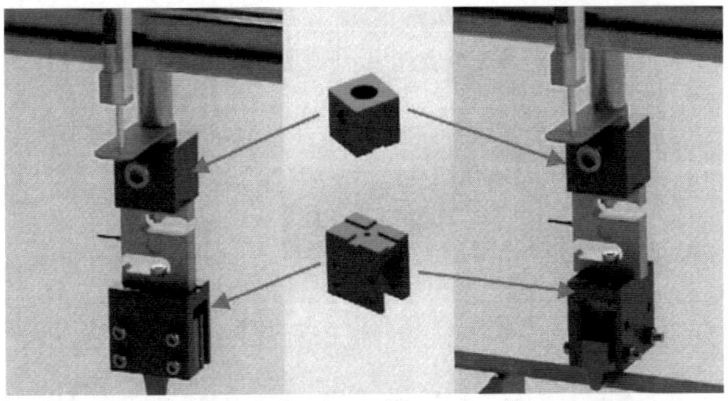

Fig. 5: The intermediate part for installing the force measurement device

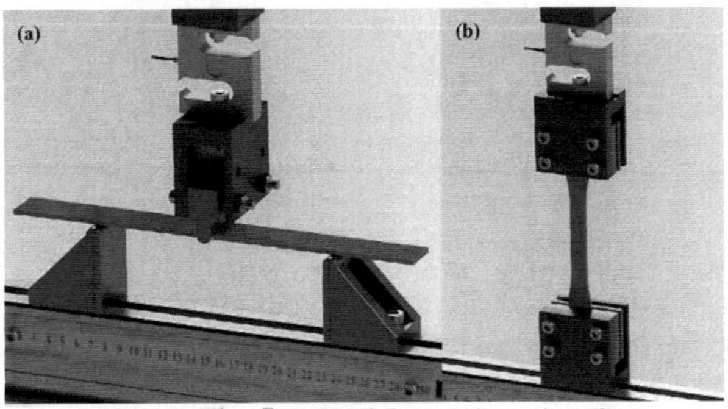

Fig. 6: The fixtures (a) and the grips (b)

IOP Conf. Series: Materials Science and Engineering **642** (2019) 012009 doi:10.1088/1757-899X/642/1/012009

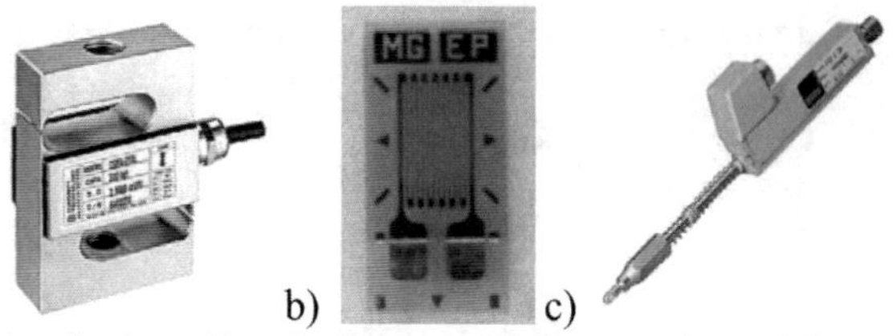

Fig. 7: The measurement devices: Loadcell (a), strain gauge (b) and linear potentiometer (c)

The final design is presented in Fig. 8.

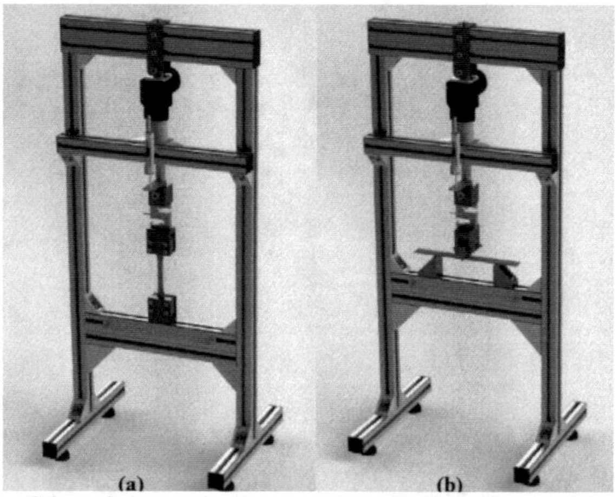

Fig. 8: Final design of the miniature testing machine for tensile test (a) and bending test (b)

2.3. Structural analysis

The sizing of components and the structural analysis are realized by finite element method using ABAQUS software.

Mechanical properties of materials are given in Table 1.

Table 1. Material properties

Materials	Young's Modulus (MPa)	Poisson ratio	Ultimate tensile strength (MPa)	Yield strength (MPa)
Shaped aluminum	69000	0.33	110	95
C45 steel	190000	0.29	630	310

2.3.1. Load and boundary conditions

The load and boundary conditions are presented correspondingly in Fig. 9 and Fig. 10.

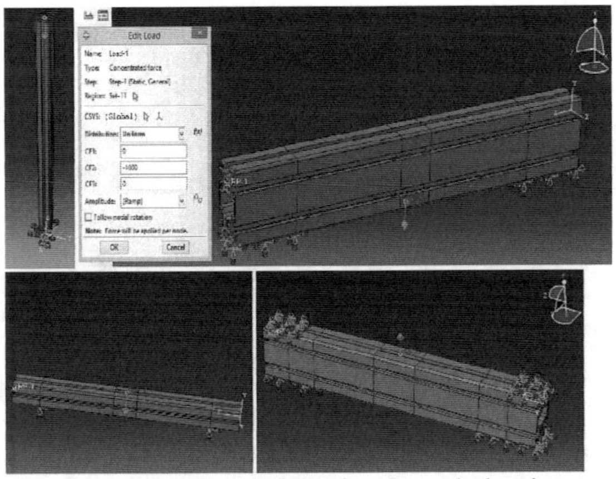

Fig. 9: Load and boundary conditions for shaped aluminum components

In fact, the real applied forces on the parts are very complex due to the influence of many factors such as the gap in the assembly and the interaction between components. However, the loading magnitude and rate are small in our test condition. So, these influencing factors will be ignored to simplify the setting of load and boundary conditions but still ensure the properties and operation principles of the components.

The frame structure can be considered as a truss. Using the theory of truss, the internal forces in the members of the truss can be calculated in case of only one external load which is the force created by the actuator during the test.

For the boundary conditions, all shaped aluminum components and intermediate parts are fixed at the position of connection.

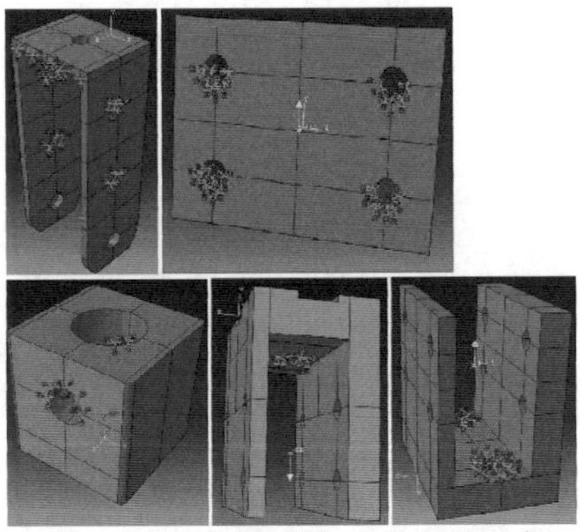

Fig. 10: Load and boundary conditions for other intermediate parts

2.3.2. Meshing

For the shaped aluminum parts, because of the complicated surface shape, the swept meshing technique is applied with element type of C3D8R (Fig. 11). In addition, the mesh will be fine in stress concentration regions and coarse in the remaining regions.

For other parts, because their shape is simpler, structured meshing technique is applied with the same element type of shaped aluminium parts (Fig. 12). The structured meshing technique generates structured meshes using simple predefined mesh topologies. Abaqus/CAE transforms the mesh of a regularly shaped region, such as a square or a cube, onto the geometry of the region needing to mesh.

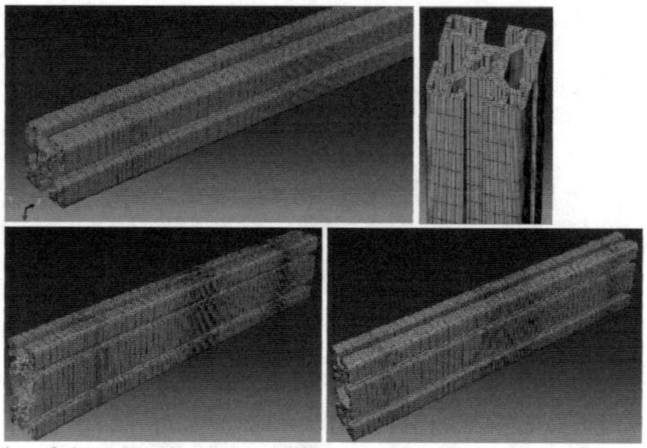

Fig. 11: Meshing models of the aluminum profile parts using bias technique with minimum size of 2 and maximum size of 10

Fig. 12: Meshing models of other parts with element size of 1

2.3.3. Simulation results

To avoid plastic deformation in the frame during loading, the maximum stresses in components do not allow to overcome the yield stress of material. To compensate for the errors from modeling simplification and mechanical properties of materials in a structural calculation, a safety factor of 3 is taken into account. Results of stress and displacement field of some parts are presented in Fig. 13 and Fig. 14. After analyzing the structure, the results show that the frame ensures durable conditions during operation: the maximum stress of all components showing in Table 2 is less more three times than the yield stress of the material. Besides, the displacement of all components is negligible (less than 0.05 mm). It means that the frame is considered solid and ensures the co-axial during a test.

Table 2: Comparison of stress to material ultimate tensile stress

Parts	$\sigma\ max\ (MPa)$	$\sigma_{YS}/3\ (MPa)$
Shaped aluminum	0.728	32
Others (C45 steel)	25.08	103

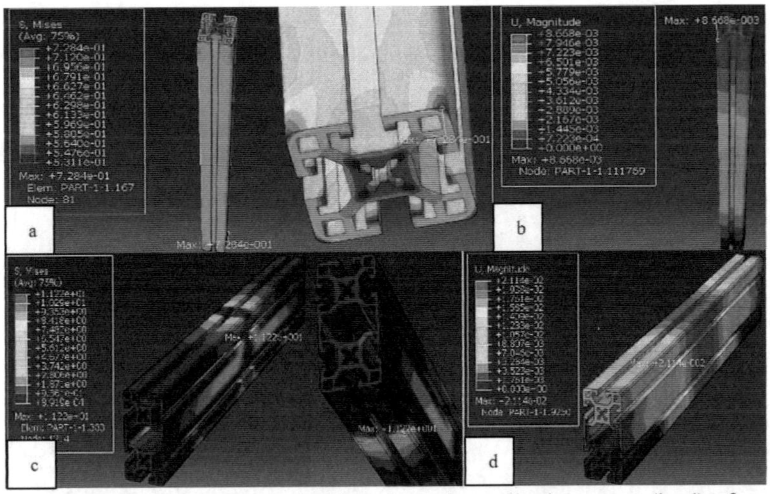

Fig. 13: Results of Von Mises stress (a, c) and resultant displacement (b, d) of some shaped aluminum parts

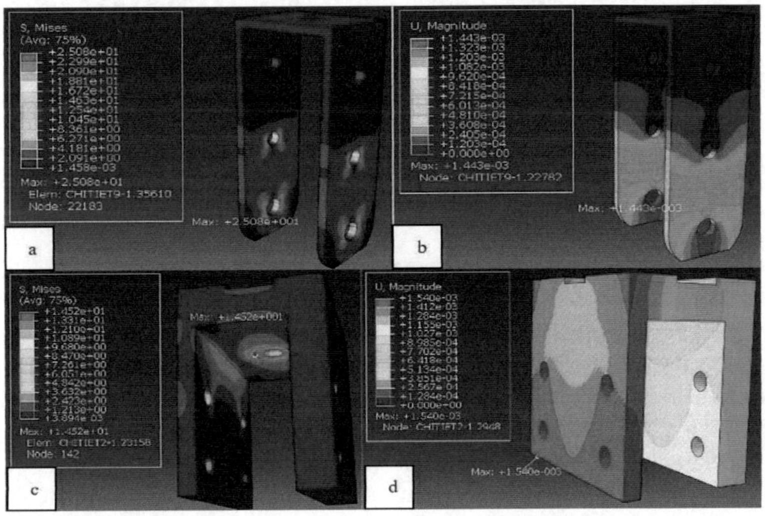

Fig. 14: Results of Von Mises stress (a, c) and resultant displacement (b, d) of some other parts

2.4. Control system

The main force of the machine is generated by a linear actuator. According to the specification of producer, maximum operating voltage of the linear actuator is 24 V corresponding to the speed of 5 mm per second. There is a limitation that the linear actuator is not able to de-assembly to set up a control system. Hence, it must be connected to a H-Bridge which transfers the signal to Arduino to control PWM. Therefore, it is possible to obtain the specific speeds satisfying the ASTM standard or the others.

In addition, a potentiometer linear 50 mm is connected to the linear actuator by mechanical connectors to transfer signals of displacement to Hocdelam USB-9090 card for the feedbacks of controlling. The connection should be ensured the co-axial during testing. It is also supposed to record the displacement in bending test.

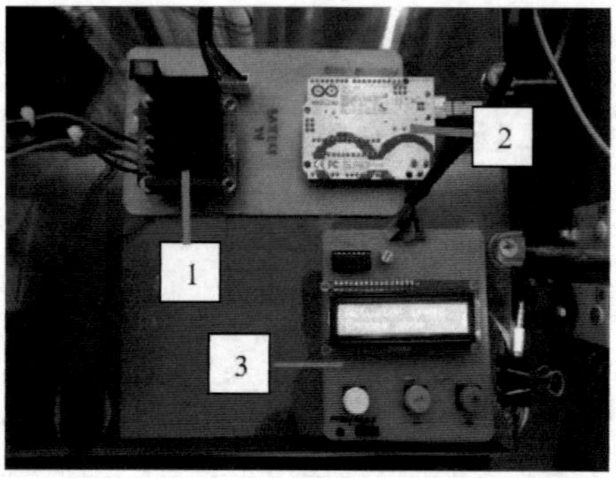

Fig. 15: Control system: 1) H Bridge; 2) Arduino; 3) Display and control button device.

2.5. Data acquisition

The data acquisition system composes a load cell, a strain gauge, two amplifiers (VAmp) for load cell and strain gauge and a Hocdelam USB-9090 card (10 bits Analog to Digital Converter) as shown in Fig. 16. The DAQ system receives the signals of output voltage from the load cell and the strain gauge, then amplifies and transforms these signals to ADC values. After calibration the sensors, these ADC values are transformed to force and strain values which are shown on computer with LabVIEW software.

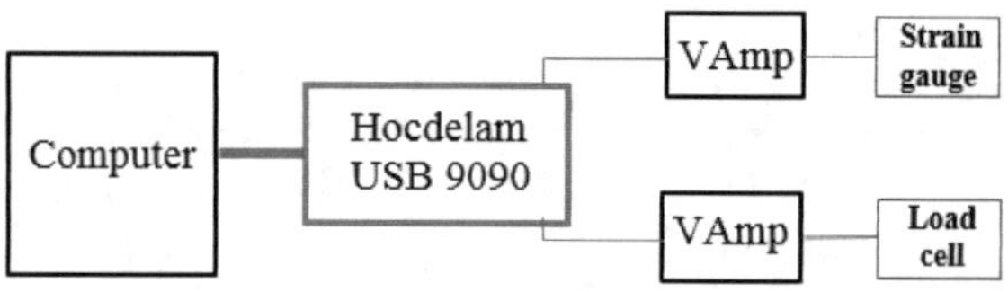

Fig. 16: The schematic of data acquisition

First of all, the data acquisition requires the signal to be stable. Due to the effect of voltage amplifier, the noise of electric signal is slightly high. To tackle these problems, it is possible to use a Kalman filter (which is not presented in this paper) in home-made LabVIEW program to reduce the noise. In addition, it is also combined with the block of sample variances to process signals smoother. Therefore, the calibration procedure in the next step is simple.

The calibration of the loadcell is realized by measuring different calibrated test weights and its corresponding output voltages as presented in Fig. 17. The output voltage is amplified and transformed to ADC value. In our system, 1 ADC is equivalent to 0.1 kg. For all different weights, the gain value is adjusted until the value of ADC corresponding to test weight with smallest errors by changing potentiometer of amplifier as shown in Fig. 18.

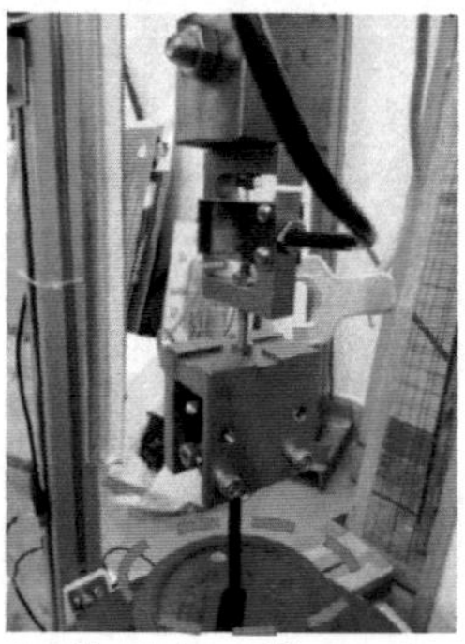

Fig. 17: Calibration of loadcell with calibrated test weight (red dash circle)

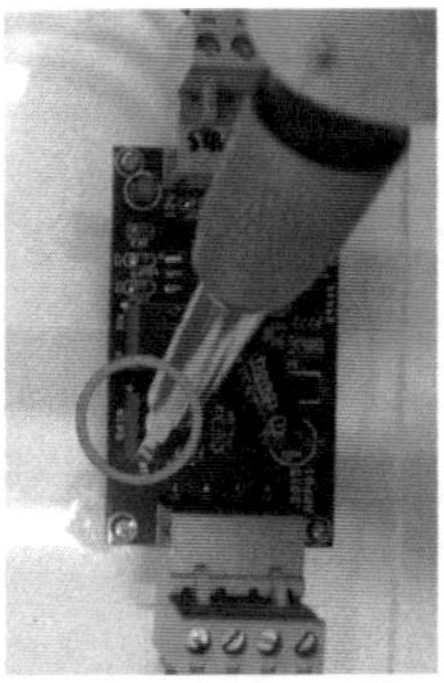

Fig. 18: Gain function of amplifier in a red circle

The results of calibration of loadcell is shown in Fig. 19. It can be seen that the ADC value increases linearly with the test weight for the chosen gain value of amplifier.

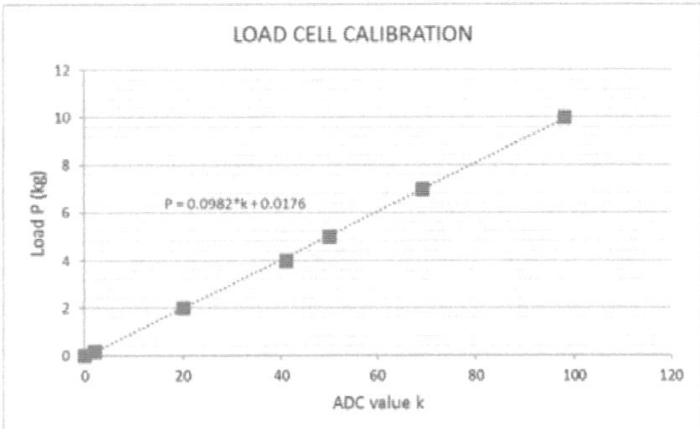

Fig. 19: Result of calibration of loadcell

In terms of strain gauge calibration, in our study, the approach method is using the cantilever beam theory with a standard steel beam having known cross-section dimension and Young's Modulus. This system composes a long beam which is fixed in one end and free in another end (Fig. 20). A dial indicator measures values of the displacement at free end when there is an applied load. The cantilever beam theory allows to define the stress state in beam. From the Young's modulus, it is possible to estimate strain at where a strain gauge is attached to the surface.

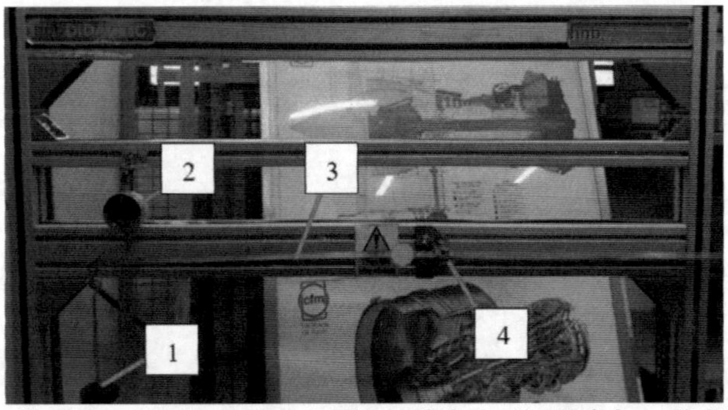

Fig. 20: System for calibrating strain gauge (at the red dot): 1) Weights; 2) Dial Indicator; 3) Steel beam; 4) Clamp

Fig. 21 presents the strain gauge calibration using cantilever beam theory. The beam is fixed at point B and is applied a load P at point A. The vertical displacement at point C is given by the dial indicator. A strain gauge is attached to the upper surface of the beam at point D. The position of these

IOP Conf. Series: Materials Science and Engineering **642** (2019) 012009 doi:10.1088/1757-899X/642/1/012009

points is defined via the distance to the fixed end B such in Fig 21. The Young's modulus of material is defined by following equation:

$$E = \frac{P}{y \times I_x}\left(\frac{z^3}{6} - \frac{L^2}{2}z + \frac{L^3}{3}\right)$$

where y is the displacement at a point far from fixed end a distance of z, I_x is area moment of inertia related to the x axis $\left(I_x = b.h^3/12\right)$ and b is width of beam.

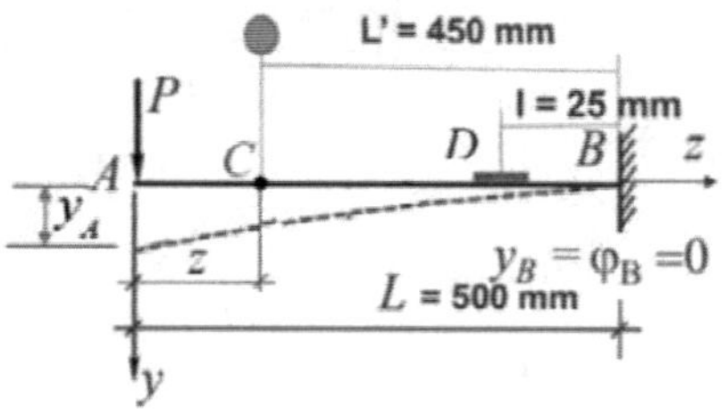

Fig. 21: The cantilever beam for strain gauge calibration

At point C, the displacement is measured by a dial indicator and the results are shown in Table 3.

Table 3: The experiment results to estimate Young's Modulus

M (kg)	*P (N)*	*y (mm)*
0	0	0
0.1	0.98066	2.428
0.2	1.96132	5.022
0.3	2.94198	7.73
0.4	3.92264	10.418
0.5	4.9033	13.018

Based on the results obtained from Table 3, evaluated Young's Modulus is 196306 MPa (1.8% difference in compared to theory value $E_{theory} = 200000$ MPa). Hence, the result is valuable.

The local stress at point D is defined in the form below:

$$\sigma = \frac{P \times l}{I_x} \times \frac{h}{2}$$

where l is the distance from the fixed end to attached strain gauge, h is the thickness of the beam.

With calculated Young's Modulus of 196306 MPa and the 30x3 mm cross-section of the beam, the strain at point D is estimated by $\varepsilon = \sigma/196306$ and shown in Table 4.

On the other hand, the output voltage of strain gauge is amplified and transformed to ADC value. To obtain the highest accuracy of strain gauge, the corresponding ADC value must be highest. It means that 1 ADC corresponds to smallest measurable strain. In the range of gain value of used amplifier, the maximum ADC value corresponding to 0.1 kg is 7. The next steps of calibration are similar to ones in loadcell calibration. The final gain value for strain gauge amplifier is obtained from measured results in Table 3 and the obtained accuracy of strain gauge is nearly 8.

Table 4: The obtained strain and ADC value corresponding

M (kg)	σ (MPa)	$\varepsilon\left(\mu m/m\right)$	ADC value
0.1	10.3514	52.7	7
0.2	20.7028	105.5	13
0.3	31.0542	158.2	19
0.4	41.4056	210.9	26
0.5	51.7571	263.6	33

The results of calibration of strain gauge is shown in Fig. 22. It can be seen that the ADC value increases linearly with the test weight for the chosen gain value of amplifier.

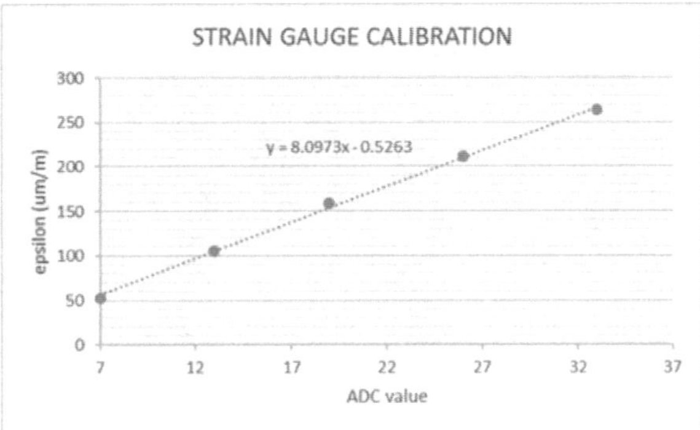

Fig. 22: Results of calibration of strain gauge

Fig. 23 depicts the GUI written in LabVIEW. It permits not only to follow the value of load and strain but also to observe their relation curve through a graph. Moreover, this GUI allow to vary the calibrated value for other sensors.

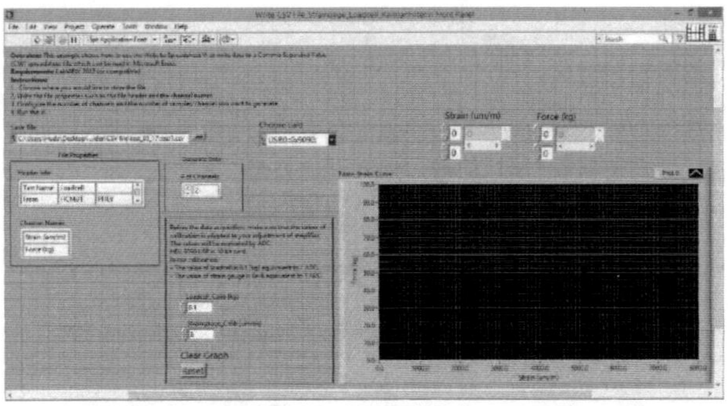

Fig.. 23: A GUI LabVIEW for tensile test

3. Results and validation

Fig. 24 shows the complete test machine under tensile test. It composes the construction, the control system for linear actuator and signal processing and acquisition system.

Fig. 24: Miniature testing machine in tensile test: 1. Control system; 2. Data acquisition system; 3. Computer

In terms of assessing the accuracy of our test machine, the strain acquisition is verified first, then some experimental results are analyzed. The strain measured directly by strain gauge is compared with one defined by the Digital Image Correlation (DIC) method (which is not presented in this paper). The results of stress-strain curve of natural fiber composites are shown in Fig. 25 and Fig. 26. It is clear that the strains measured by two different methods are equivalent. Hence, it can be concluded that the calibration of strain gauge is accurate (maximum difference is about 15%).

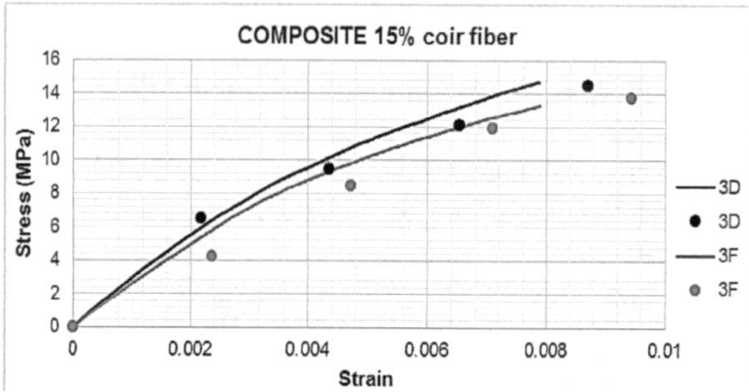

Fig. 25: Comparison of two strains on 15% coir/epoxy composite measured by DIC (dot) and strain gauge (line)

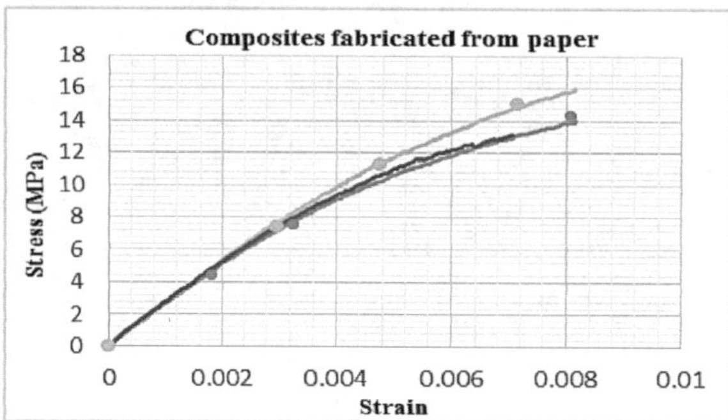

Fig. 26: Comparison of two strains on 43% paper/epoxy composite measured by DIC (dot) and strain gauge (line)

Because the composite materials in this study are made of coir fibers in form of sheet and having random orientation, they are not only an orthotropic composite as usually but also considered to be in-plane isotropic elastic material. Therefore, their shear modulus can be calculated in two different ways corresponding for two material behaviours. Table 8 shows the shear modulus of 15% fiber composite calculated from two different ways. The difference between two values of shear modulus is small (less than 5%), it can be concluded that the experimental results of 15% fiber composite group are reliable.

Table 5: Comparison of shear modulus of 15% fiber composite

Specimens	15% fiber composite
$G_{12}(MPa) = \dfrac{E}{2(1+\upsilon_{12})}$	1257
$G_{12}(MPa) = \left(\dfrac{f}{G_f} + \dfrac{1-f}{G_m}\right)^{-1}$	1218
Error (%)	3.1

From these results, it can conclude that the test machine allows to realize tests and acquire the data which is acceptable and valuable. Furthermore, the stability of machine is good with smooth curve of mechanical properties.

4. Conclusion

In this study, the design, calibration and compliance measurement of a miniature test machine for tensile test were discussed. The design has the load bearing capacity up to 3000 N. The machine permits the minimum loading rate of 1.7 mm per minute and the strain precision of 8. On the other hand, the recorded data is smooth enough to be probable to get the mechanical properties of material pretty accurate. In addition, the machine performance and compliance measurement are appreciated during tensile test. The machine has also an additional advantage that is lower cost and smaller size compared to the commercial ones. Moreover, it is ideal for the mechanical properties of the medium-strength materials. In the future, it will be supported with more accuracy card, more powerful load generator and well-being structure.

Acknowledgement

The authors would like thank to the Research Fund of Department of Aerospace Engineering, Ho Chi Minh City University of Technology for its financial support.

References

[1] R. Partherpan et al. Design and usage of a simple min-imature specimen test setup for the evaluation for mechanical properties. Inter. J. Microstructure and Materials Properties 2005; 1(1): 38-50.

[2] E. Huerta. Universal testing machine for mechanical properties of thin materials. REVISTA MEXICANA DE FÍSI-CA 2010; 4(56): 317-322.

[3] S. Cosmas Tatenda Katsambe. Effect of pulse width modulation on DC motor speed, Inter. J. Students' Research in Technology & Management 2017; 5(2): 42-45.

[4] F. Hamid. Design miniature tensile testing machine, Proceeding of 8th International Engineering Symposium 2016.